Dioxin and Dioxin-Like Compounds and Human Health

Dioxin and Dioxin-Like Compounds and Human Health

Editor

Muneko Nishijo

MDPI • Basel • Beijing • Wuhan • Barcelona • Belgrade • Manchester • Tokyo • Cluj • Tianjin

Editor
Muneko Nishijo
Kanazawa Medical
University
Japan

Editorial Office
MDPI
St. Alban-Anlage 66
4052 Basel, Switzerland

This is a reprint of articles from the Special Issue published online in the open access journal *Toxics* (ISSN 2305-6304) (available at: https://www.mdpi.com/journal/toxics/special_issues/Dioxin_Human_Health).

For citation purposes, cite each article independently as indicated on the article page online and as indicated below:

Lastname, A.A.; Lastname, B.B. Article Title. *Journal Name* **Year**, *Volume Number*, Page Range.

ISBN 978-3-0365-8040-1 (Hbk)
ISBN 978-3-0365-8041-8 (PDF)
doi.org/10.3390/books978-3-0365-8041-8

Contents

toxics

MDPI

Editorial

Dioxin and Dioxin-like Compounds and Human Health

Muneko Nishijo

Department of Public Health, Kanazawa Medial University, Uchinada 920-0293, Ishikawa, Japan; ni-koei@kanazawa-med.ac.jp

In epidemiological studies, associations of dioxin and dioxin-like (dl)-compound exposure with metabolic diseases, including diabetes and metabolic syndrome, in adults and with neurodevelopmental problems and earlier/later puberty in children have been suggested in the general population and in environmentally exposed populations.

2,3,7,8-Tetrachlorodibenzo-p-dioxin (TCDD) is the most toxic polychlorinated dibenzo-p-dioxins and -furan (PCDD/F) congener and is a by-product of the production of herbicides such as 2,4,5-trichloroacetophenoxy acetic acid (2,4,5-T). Workers in Germany, Netherland, Bohemia (former Czechoslovakia), and USA were accidentally exposed to TCDD in chemical plants producing 2,4,5-T including TCDD between the 1950s and 1970s. A large-sized population including workers and residents in Seveso, Italy was exposed to extremely high levels of TCDD caused by an industrial accident in 1976. In Missouri, USA, a significant amount of TCDD-contaminated waste oil was sprayed in residential areas in the period 1971–1972, resulting in environmental contamination of TCDD. Portland Harbor, a section of Willamette River flowing through the Willamette Valley in Oregon, USA, has been industrialized and impacted by urban and industrial activities for over a century. This area is heavily contaminated by a variety of chemicals including polychlorinated biphenyls (PCBs), PCDD/Fs, and polycyclic aromatic hydrocarbon (PAH), which exert their effect through the aryl hydrocarbon receptor (AHR).

All over southern Vietnam, the U.S. military sprayed millions of liters of herbicide such as Agent Orange, which is composed in part by 2,4,5-T, during Operation Ranch Hand (1961–1971), and not only Vietnamese residents and solders but also American and Korean soldiers were exposed to TCDD. Even after more than 40 years since the war ended, high dioxin levels, particularly high TCDD levels, have been found in the soil and sediment inside former U.S. air bases, particularly Da Nang, Phu Cat, and Bien Hoa airbases, which caused contamination of the environment and human health, including workers and residents living around the airbases.

This Special Issue includes a total of 10 articles, including 6 original articles and 3 reviews, aiming to provide recent or overall study results to investigate the effects of dioxins on human health, focusing on children and adults exposed to historical and/or present pollution in Vietnam. One more original article in the present Special Issue is a study showing that coupling environmental whole mixture toxicity screening with unbiased RNA-seq using biological responses in Danio rerio (zebrafish) is a useful method to evaluate the toxicity of a mixture of compounds with AHR, such as samples at numerous sites in the Portland Harbor Superfund Site.

Articles by Pham-The et al. [1] and Tran et al. [2] in this issue were results of follow-up studies of children from the Da Nang cohort in Vietnam. Da Nang airbase, located in central Vietnam, is one of a number of former U.S. airbases contaminated with dioxins as a result of the use of Agent Orange and other herbicides. In 2007–2008, we previously measured levels of 17 PCDD/F congeners in the breast milk of mothers residing nearby Da Nang airbase; significantly, we found that these levels were 3–4 times higher than those of mothers living in unsprayed areas, suggesting that dioxin exposure originating from herbicide is still high enough to increase health risks in the residents living in nearby areas of Da Nang airbase.

Citation: Nishijo, M. Dioxin and Dioxin-like Compounds and Human Health. *Toxics* **2023**, *11*, 512. https://doi.org/10.3390/toxics11060512

Received: 30 May 2023
Accepted: 2 June 2023
Published: 6 June 2023

Follow-up studies of this Da Nang birth cohort identified adverse effects of dioxin exposure on infant and child neurodevelopment from 4 months to 8 years of age, including increased autistic traits (poor social and communication abilities) associated with high TCDD exposure and poor language and motor development associated with high TEQ-PCDD/F exposure in boys at 3 years of age. Boys showed poorer cognitive ability associated with high TCDD exposure and poor coordination movement skills associated with TEQ-PCDD/Fs exposure at 5 years of age.

Regarding ADHD likelihood, Pham-The et al. [1] reported that hyperactivity scores, a subscale of ADHD rating scale, were significantly higher in boys with high TCDD at 5 years of age, but no association between ADHD symptoms and dioxin exposure was found in girls. In contrast, at 8 years of age, girls showed high hyperactivity scores in the high-TCDD group, which was also significantly associated with unusual behavior scores according to the Autism Spectrum Rating Scale (ASRS), suggesting high perinatal TCDD exposure may increase ADHD likelihood and autistic traits, particularly in girls.

At 8 years of age, boys with high TCDD showed increased reading learning difficulties, although no increase in ADHD symptoms was found. Tran et al. [2] summarized all study results from 4 months to 8 years of age and concluded that perinatal TCDD exposure impacts social–emotional cognitive functions, leading to sex-specific neurodevelopmental disorders, i.e., learning difficulty in boys and ADHD in girls.

Manh et al. [3] collected 45 blood samples from 9-year-old children living in areas nearby Phu Cat airbase, one of the former U.S. airbases, and 35 blood samples of 9-year-old children in the nonexposed area to make 12 pooled samples, and they measured 17 PCDD/F congeners in sera. The mean of the TEQ of PCDD/Fs level in Phu Cat was more than three times higher than that in the nonexposed area, but no TCDD was detected even in Phu Cat samples. The serum levels of some congeners, but not TEQ-PCDD/Fs, were correlated with those in breast milk, which were collected from mothers when they were nursing their children.

Pham N.T. et al. [4] investigated the effect of perinatal dioxin exposure (indicated by dioxins in breast milk) on the gaze behavior of 142 children at 3 years of age from the 2012 Bien Hoa birth cohort; the children, residing in the areas most contaminated areas by TCDD, originating from Agent Orange around Bien Hoa airbase in Vietnam, were examined via an eye-tracker. The gaze fixation duration on facial areas when viewing 10 still images of children was calculated as the gaze behavior index. The face fixation duration (%) significantly decreased as TCDD concentrations increased in a dose–effect manner in girls, which suggested atypical gaze behavior for watching human faces. Furthermore, these girls with atypical gaze behavior showed lower social communication scores according to the ASRS and smaller head sizes, suggesting increased autistic traits in girls.

Between 2007 and 2015, Nishijo et al. [5] collected a total of 861 breast milk samples (597 samples from 3 herbicide-sprayed areas including Quang Tri, Da Nang, and Bien Hoa in southern Vietnam; 264 samples from 3 unsprayed areas in northern Vietnam) and determined PCDD/Fs in each sample from mothers 1 month after delivery. The levels of TEQ-PCDD/Fs and 17 PCDD/F congeners were significantly higher in the sprayed area samples than the unsprayed area samples. The authors found particular PCDD/F congener patterns for different areas. High TCDD concentrations were found in Bien Hoa, high levels of TCDD and 1,2,3,6,7,8-hexaCDD were found in Da Nang, and high 1,2,3,4,6,7,8-heptaCDD levels were found in Quan Tri. High 1,2,3,4,7,8-hexaCDF and 1,2,3,4,6,7,8-heptaCDF concentrations were also found in Da Nang and Quang Tri. The associations between the levels of TCDD, 1,2,3,4,7,8-HexaCDF, and 1,2,3,4,6,7,8-HeptaCDF were different for samples from primipara and multipara mothers, suggesting that breast feeding affected PCDF concentrations more than PCDD concentrations.

Vu et al. [6] investigated associations between dioxin exposure and brain structural irregularities in 32 Vietnamese men living near Bien Hoa airbase. Two exposure markers were used: 1) blood dioxin levels as a marker of exposure in adulthood, and 2) perinatal dioxin exposure during pregnancy, estimated by a maternal residency in the areas around Bien Hoa airbase. All subjects underwent brain magnetic resonance imaging (MRI) scans. Correlations between regional grey matter volumes and blood dioxin levels and between brain regional volumes of men with and without perinatal dioxin exposure were determined by voxel-based morphometry (VBM). Blood TCDD was associated with a low volume of the medial temporal pole and fusiform gyrus. Levels of TEQ-PCDDs were correlated with low medial temporal pole volume. However, 1,2,3,4,7,8-HxCDD was associated with high middle frontal gyrus and cerebellum volume. In men with perinatal dioxin exposure, the left inferior frontal gyrus pars orbitalis volume was significantly lower than in those without perinatal exposure. These results suggest that dioxin exposure during the perinatal period and in adulthood may cause altered regional brain volume, which can lead to cognitive deficits and unusual social–emotional behavior.

Pham P.Q. et al. [7] collected liver biopsy samples for histopathological examination from 33 chronic hepatitis patients living around the Da Nang Airbase. They found that increased TCDD levels in blood were associated with increased levels of liver function markers such as aspartate aminotransferase (AST), alanine aminotransferase (ALT), protein and total bilirubin, and high liver fibrosis stages classified using the METAVIR fibrosis staging system for histopathological examination. Similarly, increased TEQ-PCDD/Fs levels were associated with higher levels of AST and protein and the liver fibrosis stage. These findings suggest TCDD exposure may influence liver cells to increase fibrosis leading to an increased risk of liver cancer, suggesting that regular health check-ups, particularly liver function tests and imaging examinations, should be required for all subjects living in dioxin contamination areas in Vietnam.

Takiguchi et al. [8] reviewed previous publications in which the effects of PCDD/Fs and dioxin-like PCBs on the teeth and bones of animals and humans were found to identify future research directions, particularly for epidemiological studies of populations exposed to PCDD/Fs in the environment. Previously, it has been reported that exposure of fetuses to PCDD/Fs may affect odontogenesis, particularly enamel formation in human and animals. However, the effects of PCDD/Fs on bone genesis are limited to palatine bone. Exposure to PCDD/Fs during infancy may affect both teeth and bones, but the effects on bones may be reversible. High PCDD/Fs exposure even during adulthood may adversely affect teeth in human and animals. In contrast, however, PCDD/Fs exposure may induce osteogenesis and improve bone properties because the disrupting effects of PCDD/Fs cause bone remodeling and vitamin D activation in animals. More studies involving humans are required to investigate previously found associations between the PCDD/F concentrations and biological markers for teeth and bones, including metabolites of vitamin D.

The aim of the review by Vuong [9] is to discover whether there is a relationship between dioxin exposure and cancer incidence in the hotspot regions of Vietnam by estimating the risk ratio index. The results of the study show that the incidence of cancer (soft tissue sarcoma; Hodgkin's and non-Hodgkin's lymphoma; lung, prostate, and liver cancer) in the dioxin-exposed Vietnamese population is much higher than that found in the results of studies published in other countries, as a result of the high levels of dioxins in southern Vietnam, where Agent Orange was sprayed during the Vietnam War. Further studies on the health effects of dioxins in the Vietnamese population, including cancer incidence, should be conducted with improved research methods.

Rude et al. [10] analyzed the toxicity of passive sampling device extracts from two points (river mile 6.5 W and 7 W) in the Portland Harbor Superfund Site in Oregon, USA, which has been industrialized and heavily contaminated by a variety of chemicals including PCBs, PCDD/Fs, and PAH, using coupling environmental whole mixture toxicity screening with unbiased RNA-Seq. The toxicity was evaluated using developmental toxicity assays in Danio rerio (zebrafish) shown as "wavy" notochord malformation. The gene expression from two extracts was parallel, although it was more evident in river mile 6.5 W than 7.0 W. Differential expression, reminiscent of the wavy notched phenotype, was not accounted for by either class of chemicals. Therefore, these techniques offer a compelling method for non-targeted hazard characterization of whole mixtures without requiring complete chemical characterization.

Acknowledgments: We would like to thank all medical doctors and nurses in commune heath stations in all study areas, district hospitals in Da Nang, Dong Nai Prefectural Hospital in Bien Hoa, and 103 Military Hospital in Da Dong, Hanoi, in Vietnam. We also thank Ryumon Honda and Kenji Tawara for their technical support to analyze dioxins in breast milk.

Conflicts of Interest: The authors declare no conflict of interest. The funders had no role in the design of the study; in the collection, analyses, or interpretation of data; in the writing of the manuscript; or in the decision to publish the results.

References

1. Pham-The, T.; Nishijo, M.; Pham, T.N.; Vu, H.T.; Tran, N.N.; Tran, A.H.; Hoang, L.V.; Do, Q.; Nishino, Y.; Nishijo, H. Perinatal dioxin exposure and attention deficit hyperactivity disorder (ADHD) symptoms in children living in a dioxin contamination hotspot in Vietnam. *Toxics* **2022**, *10*, 212. [CrossRef] [PubMed]
2. Tran, N.N.; Pham-The, T.; Pham, T.N.; Vu, H.T.; Luong, K.N.; Nishijo, M. Neurodevelopmental effects of perinatal TCDD exposure differ from those of other PCDD/Fs in Vietnamese children living near the former us air base in Da Nang, Vietnam. *Toxics* **2023**, *11*, 103. [CrossRef] [PubMed]
3. Manh, H.D.; Kido, T.; Takasuga, T.; Yamashita, M.; Giang, L.M.; Nakagawa, H. The relationship of dioxin levels in serum of 9-year-old Vietnamese children and their mothers' breast milk. *Toxics* **2022**, *10*, 155. [CrossRef] [PubMed]
4. Pham, T.N.; Nishijo, M.; Pham-The, T.; Tran, N.N.; Vu, H.T.; Tran, A.H.; Tran, T.V.; Nishino, Y.; Nishijo, H. Effect of perinatal dioxin exposure originating from Agent Orange on gaze behavior in 3-year-old children living in the most dioxin-contaminated areas in Vietnam. *Toxics* **2022**, *10*, 150. [CrossRef] [PubMed]
5. Nishijo, M.; Vu, H.T.; Pham-The, T.; Pham, T.N.; Tran, N.N.; Nakagawa, H.; Nishijo, H. Dioxin congener patterns in breast milk samples from areas sprayed with herbicide during the Vietnam War 40 Years after the War Ended. *Toxics* **2022**, *10*, 323. [CrossRef] [PubMed]
6. Vu, H.T.; Pham, T.N.; Yokawa, T.; Nishijo, M.; Pham-The, T.; Do, Q.; Nishino, Y.; Nishijo, H. Alterations in regional brain regional volume associated with dioxin exposure in men living in the most dioxin-contaminated area in Vietnam: Magnetic resonance imaging (MRI) analysis using voxel-based morphometry (VBM). *Toxics* **2021**, *9*, 353. [CrossRef] [PubMed]
7. Pham, P.Q.; Nguyen, V.B.; Pham, T.T.; Duong, N.X.; Nguyen, H.T.; Ha, Q.V.; Nguyen, T.D.; Hoang, T.M.; Dinh, D.T.; Tran, Q.T.N.; et al. Histopathological alterations in the livers of chronic hepatitis patients exposed to Agent Orange/dioxin in Vietnam. *Toxics* **2022**, *10*, 315. [CrossRef] [PubMed]
8. Takiguchi, T.; Vu, H.T.; Nishino, Y. Effects of polychlorinated dibenzo-p-dioxins, polychlorinated dibenzofurans, and dioxin-like PCBs on teeth and bones in animals and humans. *Toxics* **2022**, *11*, 7. [CrossRef] [PubMed]
9. Vuong, T.P. Research on the relationship between exposure to the dioxins and cancer incidence In Vietnam. *Toxics* **2022**, *10*, 384. [CrossRef] [PubMed]
10. Rude, C.I.; Tidwell, L.G.; Tilton, S.C.; Waters, K.M.; Anderson, K.A.; Tanguay, R.L. Coupling environmental whole mixture toxicity screening with unbiased RNA-Seq reveals site-specific biological responses in Zebrafish. *Toxics* **2023**, *11*, 201. [CrossRef] [PubMed]

toxics

MDPI

Article

Coupling Environmental Whole Mixture Toxicity Screening with Unbiased RNA-Seq Reveals Site-Specific Biological Responses in Zebrafish

Christian I. Rude [1], Lane G. Tidwell [1], Susan C. Tilton [1], Katrina M. Waters [1,2], Kim A. Anderson [1] and Robyn L. Tanguay [1,*]

[1] Department of Environmental and Molecular Toxicology, Oregon State University, Corvallis, OR 97331, USA
[2] Pacific Northwest National Laboratory, Biological Sciences Division, Richland, WA 99354, USA
* Correspondence: robyn.tanguay@oregonstate.edu; Tel.: +1-(541)-737-3608

Abstract: Passive sampling device (PSD) extracts paired with developmental toxicity assays in *Danio Rerio* (zebrafish) are excellent sensors for whole mixture toxicity associated with the bioavailable non-polar organics at environmental sites. We expand this concept by incorporating RNA-Seq in 48-h post fertilization zebrafish statically exposed to PSD extracts from two Portland Harbor Superfund Site locations: river mile 6.5W (RM 6.5W) and river mile 7W (RM 7W). RM 6.5W contained higher concentrations of polycyclic aromatic hydrocarbons (PAHs), but the diagnostic ratios of both extracts indicated similar PAH sourcing and composition. Developmental screens determined RM 6.5W to be more toxic with the most sensitive endpoint being a "wavy" notochord malformation. Differential gene expression from exposure to both extracts was largely parallel, although more pronounced for RM 6.5W. When compared to the gene expression associated with individual chemical exposures, PSD extracts produced some gene signatures parallel to PAHs but were more closely matched by oxygenated-PAHs. Additionally, differential expression, reminiscent of the wavy notochord phenotype, was not accounted for by either class of chemical, indicating the potential of other contaminants driving mixture toxicity. These techniques offer a compelling method for non-targeted hazard characterization of whole mixtures in an in vivo vertebrate system without requiring complete chemical characterization.

Keywords: passive sampling; *Danio rerio*; polycyclic aromatic hydrocarbons; mixtures; developmental; transcriptomics

Citation: Rude, C.I.; Tidwell, L.G.; Tilton, S.C.; Waters, K.M.; Anderson, K.A.; Tanguay, R.L. Coupling Environmental Whole Mixture Toxicity Screening with Unbiased RNA-Seq Reveals Site-Specific Biological Responses in Zebrafish. *Toxics* **2023**, *11*, 201. https://doi.org/10.3390/toxics11030201

Academic Editor: Muneko Nishijo

Received: 25 January 2023
Revised: 19 February 2023
Accepted: 20 February 2023
Published: 21 February 2023

1. Introduction

The Portland Harbor Superfund Site (PHSS) was designated by the U.S. EPA in 2000 along approximately nine miles of the Willamette River flowing through Portland, Oregon due to the widespread contamination of sediments by chemicals including polychlorinated bisphenols (PCBs), dioxins, DDT degradation products, and polycyclic aromatic hydrocarbons (PAHs) [1]. Since then, the area has undergone significant remediation including dredging and capping at particularly contaminated sites [1]. Environmental monitoring and remediation efforts continue to this day and are expected to continue at least into the late 2020s [2].

The Oregon State University Superfund Research Program pioneered a technique utilizing low density polyethylene passive sampling devises (LDPE-PSDs) and developmental toxicity assays in *Danio rerio* (zebrafish) to measure toxicant abundance and bioactivity at numerous sites of interest within the PHSS [3,4]. Dechorionated developing zebrafish provide an excellent vertebrate biosensor of toxicity because they are easily reared, high throughput, and exquisitely sensitive to insult by bioactive compounds [5]. PSDs collect a time weighted average of the freely dissolved and therefore bioavailable fraction of contaminants at a site [6,7]. Extracts from PSDs yield whole environmental mixtures for chemical and toxicological characterization. The technique uses PSDs deployed in pairs. One PSD is spiked with performance compounds for more precise chemical measurement while the other remains unspiked to prevent interference in the bioassay. This technique was termed as Biological Response Indicators Devices Gauging Environmental Stressors (BRIDGES). Using it, Allan et al. found a significant correlation between the PAH concentrations and associated site toxicities in the PHSS [3].

Sediment within the PHSS contains high levels of PAHs, which in turn partition into the river water [1,8]. While not all PAHs appear to be toxic, many such as the infamous benzo(a)pyrene are classified as known or suspected human carcinogens, and others cause toxicity to the immune, cardiovascular, and neurological systems [9–16]. Canonically, many PAHs exert their effect through the aryl hydrocarbon receptor (AHR), a transcription factor that resides in the cytosol until a properly fitting ligand shifts its conformation [17]. It then breaks free from its chaperones and enters the nucleus where it canonically partners with AHR nuclear translocator (ARNT) to bind xenobiotic response elements within the genome. This induces the transcription of numerous genes involved in the xenobiotic response including p450s, wfkkn1n, foxq1a, and nrf2 [18–22].

PAHs are present in the environment as complex mixtures [23], however, traditional toxicity testing evaluates PAHs on an individual basis. BRIDGES is an ideal tool because it allows for toxicological evaluation of whole mixtures, even in the absence of complete chemical data. This study expands the concept of BRIDGES by incorporating RNA sequencing to interpret the molecular responses behind site specific toxicity in developing zebrafish. We set out to characterize the hazard of environmental mixtures, compare the molecular responses between PSD extract exposures, and determine how the molecular response to mixtures informs the drivers of toxicity.

2. Materials and Methods

2.1. Chemicals

Dimethylsulfoxide (DMSO), dichloromethane (DCM), hydrochloric acid (HCL), n-hexane, and isopropanol used for PSD precleaning and cleaning and chemical sample prep were Optima grade or better and purchased from Fisher Scientific. Performance reference compounds (PRCs) included perdeuterated flourene-D10, p,p′-DDE-D8, and benzo[b]fluoranthene-D10. Perdeuterated surrogate recovery standards included naphthalene-D8, acenaphthylene-D8, phenanthrene-D10, fluoranthene-D10, pyrene-D10, benzo[a]pyrene-D12, and benzo[ghi]perylene-D12. Samples were spiked with perylene-D12 before GCMS/MS analysis as an internal standard. The 33 PAHs included in the quantitative PAH method and their abbreviations are included in Supplementary Table S1.

2.2. Sampling Methods

Lipid free PSDs were made from 2.7 × 100 cm LDPE tubing that was pre-cleaned with n-hexane and heat sealed at both ends, as described in Sower et al. [24]. PSDs intended for chemical analysis were spiked with deuterated reference compounds, while those intended for toxicity testing were not. At each sampling site, stainless steel cages containing five PSDs were deployed in matched pairs with one member of the pair intended for chemical analysis and the other intended for toxicity testing. Cages were secured ~3 m above the river floor with the anchored floatation setup described in Anderson et al. [25]. Each deployment lasted 30 days, at which point the PSDs were collected, sealed in amber glass jars, transported back to the lab in coolers, cleaned with HCL and isopropanol, and stored at −20 °C until extraction.

The original sampling campaign has been described in detail by Allan et al. (2012) [3]. PSDs were deployed at nine locations: six within the Portland Harbor Superfund Site, two in the Columbia River, and one in the Willamette River above the Superfund Site. Sampling was conducted during the months of September and October 2009, and July–October in 2010 for a total of six different sampling periods. This study utilized extracts at river mile 7 West from September 2009 and river mile 6.5 West from July 2010 for chemical analysis and toxicological characterization. The extracts are referred to as RM 7W and RM 6.5W, respectively, throughout this study.

The set of five PSDs from each cage were extracted together to yield one environmental extract per cage. The PSDs from cages intended for chemical analysis were spiked with perdeuterated PAH recovery standards before extractions to account for any losses. Extractions were carried out via two sequential dialyses in 40 mL of n-hexane per PSD. The first dialysis was 4 h long while the second was 2 h. The resulting 400 mL of dialysates were quantitatively concentrated to 1 mL before chemical analysis. The dialysates intended for toxicological testing were quantitatively solvent exchanged to dimethylsulfoxide (DMSO).

2.3. Chemical Analysis

All chemical analyses were carried out on an Agilent 5975B gas chromatograph-mass spectrometer (GC-MS) equipped with a DB-5MS column (30 m × 0.25 mm, 0.25 µm).

PAH quantitation was accomplished in electron impact mode (70 EV) using selective ion monitoring. The GC injection port was held at 300 °C and the helium flow was 1.0 mL min^{-1} throughout. The oven temperature was as follows: initial 1 min hold at 70 °C, ramp rate of 10 °C min^{-1} to 300 °C, hold 4 min, ramp rate of 10 °C to 310 °C, hold 4 min. The MS quadrupole, source, and transfer line were held at 150 °C, 230 °C, and 280 °C, respectively. Each PAH was quantitated by the relative response to deuterated surrogate standards on a nine-point calibration curve with a minimum correlation coefficient of 0.98. Deuterated standard recoveries were between 44 and 109%. Lower recoveries occurred in the lower molecular weight deuterated standards due to volatile loss especially during the sample concentration. Target PAHs were recovery corrected based on the volatile loss of the deuterated recovery surrogates. The measured concentrations for each sample determined by the 33 PAH method are available in Supplementary Data S1.

The minimum detection limit for any of the PAH analytes was 10 pg/uL. Calibration controls were run every ten samples and were required to verify concentration within 15% of the true concentration to pass. All measurements reported in this study were preceded and followed by passing the calibration controls. Triplicate measurements from RM 7W were used to estimate the variance in both reported samples, as demonstrated by Matzke et al. [26].

The 1201 chemical screen was accomplished with the GC-MS running on full scan acquisition mode (mass range 5–50) and absolute EMV mode. The helium flow was 2.3 mL min^{-1} throughout. The oven temperature was as follows: initial 2 min hold at 70 °C, ramp rate 25 °C min^{-1} to 150 °C, ramp rate of 3 °C min^{-1} to 200 °C, ramp rate of 8 °C min^{-1} to 280 °C, hold 15 min, ramp rate of 40 °C min^{-1} to 310 °C, hold 3 min. The retention time was locked on chlorpyrifos at 19.23 min. The MS component temperatures were the same as the PAH method.

The 1201 chemical screen included chemicals of concern from a broad array of chemical classes including, but not limited to, chlorinated bisphenols, parent and substituted PAHs, pharmaceuticals, phthalates, and synthetic musks. Detected chemicals were identified via their mass spectra using Agilent Deconvolution Reporting Software developed by Agilent utilizing the AMDIS spectral database [27]. The library of potential chemicals was limited to GC-MS compatible compounds with some degree of hydrophobicity, as would be required for sequestration to the LDPE-PSDs.

Beyond the calibration controls, the quality control measures included laboratory preparation blanks, field and trip blanks corresponding to every deployment, laboratory cleanup blanks, and reagent blanks that accounted for approximately one third of all samples. All target compounds were below the LOD in the quality control samples.

2.4. Zebrafish Rearing, Exposure, and Morphological Assessment

Embryos used in the extract exposures were reared at the Sinnhuber Aquatic Research Laboratory from pathogen free tropical 5D strain zebrafish maintained according to the appropriate Institutional Animal Care and Use Committee (IACUC) approvals. Parent fish were housed in 100 gal tanks on a recirculating water system under 14–10 h light–dark cycle and fed twice a day with Gemma Micro 300. Water was maintained at 28 °C with a conductivity of 1000 µSiemens. Parent fish were group-spawned in the morning with the start of the light cycle. The resulting embryos were inspected for malformation, grouped by developmental stage, and enzymatically dechorionated at 4 h post fertilization (hpf). At 6 hpf, the embryos were transferred to a 100 µL exposure solution in 96-well plates, accessed again for viability, sealed, and incubated in the dark at 28 °C.

For the morphologic toxicity assessments, developing zebrafish were statically exposed in E2 embryo medium (EM) with 1.00%, 0.20%, 0.04%, and 0.008% PSD extract concentrations. The 100% PSD extract in DMSO became a 100× PSD stock, while the 20×, 4×, and 0.8× stocks were made by serial dilution in DMSO. Exposure solutions were made by a 1:100 addition of PSD stock into EM, resulting in 1% DMSO solutions. The most concentrated exposure corresponded to ~1000 times the contaminate levels in the corresponding river water. Each concentration of PSD extract was tested in 40 fish in two separate plates. Each plate contained 20 embryos exposed at one of the four concentrations of PSD extracts, eight negative controls exposed to 1% DMSO, and eight positive controls exposed to trimethyltin at 5 µM. At 5 µM, trimethyltin elicited morphological malformation in 100% of embryos, but less than 20% mortality at 120 hpf. Plates failed quality control if mortality exceeded two fish in either of the controls. Failure of either the positive or negative control, or inconsistency between the two exposure plates, resulted in a retest.

At 24 hpf, the developing zebrafish were assessed for mortality, delayed development, and notochord malformations. Then, at 120 hpf, the larvae were evaluated for mortality, abnormal touch response, yolk sac edema and pericardial edema, craniofacial abnormalities, and malformations of the body axis, brain, lower trunk, muscles, skin, and notochord. A wavy notochord malformation was also observed and included as part of the notochord endpoint. The dose response data for each exposure concentration is included in Supplementary Data S2.

2.5. RNA Isolation

The transcriptomic data generated from zebrafish larval exposures to PSD extracts were new to this study while that of the oxygenated PAH (OPAH) and parent PAH compounds have previously been published [21,28]. A summary table of the exposure scenarios and techniques to generate the RNA-Seq data is included in Supplementary Table S2. The PHSS extract and OPAH RNA samples were generated via the batch static, waterborne exposure of 20 embryos in 2 mL of solution in amber glass vials. For each PHSS sample, a ×75 PSD stock solution was diluted 1:100 in EM to yield a 0.75% PSD exposure solution. Embryos were exposed to phenanthraquinone (PHEQ) at 1.2 µM. Benz(a)anthracene quinone (BAAQ) and benzanthrone (BEZO) were used at exposure concentrations of 10 µM, which corresponded to the concentration producing a 100% effect at 120 hpf but no effect at 48 hpf. We also generated vehicle controls by batch exposing embryos to 1% DMSO. For each treatment, three groups of 20 fish were homogenized in RNAzol (Molecular Research Center, Inc.), using 0.5 mm zirconium oxide beads (Next Advance) in a bullet blender. Samples were stored at −80 °C until phenol guanidine extraction. The RNA concentration was determined with a Synergy MX microplate reader. The quality was determined with an Agilent Bioanalyzer 2100. We confirmed a RIN score above 9 in each sample before sending the total RNA samples to The University of Oregon Genomics Core, where RNA was poly-A selected via the Dynabead mRNA Purification Kit (Invitrogen), library prepped with the ScriptSec v2 Kit and ScriptSeq index primers, and 50 bp paired-end sequenced with an Illumina HISEQ 2000 sequencer.

This study also utilized the transcriptomic data measured from larval exposures to the parent PAHs retene (Ret), benzo(k)fluoranthene (BkF), and benzo(b)fluoranthene (BbF) originally generated and published by Shankar et al. [21]. We extracted the raw fastqs of these exposures from the GEO database, ascension number GSE171944. The exact generation of mRNA measured in these files is detailed in the aforementioned study, but briefly: dechorionated zebrafish were individually exposed to Ret, BkF, BbF, or 1% DMSO vehicle control in 96-well plates from 6 to 48 hpf. The concentrations used in the RNA generation exposures were equivalent to those that elicited 80% embryos exposed from 6 to 120 hpf. We refer to these concentrations as the EC80. At 48 hpf, fish were pooled into groups of eight, homogenized in RNAzole with 0.5 mm zirconium beads in a bullet blender, and purified with the Direct-zol Miniprep Kit including the optional in-column DNASE digestion step. After verifying the RNA purity and quality, the total RNA was sent to Oregon State University's Center for Genome Research and Biocomputing for poly-A enrichment, library prep, and 100 bp multiplexed sequencing on an Ilumina Hiseq 3000.

2.6. Sequencing Data Pipeline

Beginning with raw fastqs, reads were accessed for quality using FASTQC [29]. Reads were not trimmed because the initial read trimming trials did not show improvement mapping to the genome. Fastqs were mapped to the GRZ11 genome using STAR Aligner, and reads were tallied with HTSEQ [30,31]. Further analysis was accomplished in R [32]. Unless otherwise stated, graphs were made using ggplot2. We used multidimensional scaling of the raw reads to identify outliers within the treatment groups, which resulted in the exclusion of a single treatment from the treatment groups RM 6.5W, BbF, and one vehicle control. Each of these groups was reduced to n = 2, while the remaining treatment groups analyzed in this study had n = 3 or n = 4 samples. Differential expression was determined with DESEQ2 using all treatment groups to estimate common dispersion [33]. The resulting log_2foldchange (Log_2FCs) and Benjamin–Hochber adjusted *p*-values (p_{adj}) were used to determine the differentially expressed genes (DEGs) for further analysis. All DEGs with determined during analysis are included with thier log_2FC and p_{adj} values in Supplementary Table S3.

2.7. RNA-Seq Analysis

All heatmaps were made with pheatmap in R [32]. The gene expression heatmaps showed the log_2FC of every gene defined as a DEG ($|log_2FC| \geq 1$, $p_{adj} \leq 0.05$) in at least one of the displayed treatments. In the direct PHSS extract comparison, rows were clustered by Euclidian distance. In the comparison with single chemical treatments, the rows were clustered via k-means clustering with k = 8 and 500 starts. Gene Ontology analysis was performed using g:profiler2 in R. We used the Gene Ontology (GO) databases for Biological Processes (BP), Cellular Components (CC), and Molecular Function (MF) [34,35]. Databases were filtered to GO terms that included 15–400 genes to remove overly broad or narrow terms. We broadened our analysis by reducing the $|log_2FC|$ cutoff for DEGs to 0.5. The GO terms were tested for overrepresentation by a hypergeometric test and corrected for multiple tests using the built in "g_SCS" functionality. Network analysis of the GO terms was performed using the Cytoscape app Enrichment Map clustering GO terms by shared gene sets [36,37]. To generate the GO term heatmap, significantly enriched GO terms were ranked by the enrichment ratio within each treatment and database. The top 60 GO terms from each treatment were then clustered hierarchically by Jaccard distance. GO terms with Jaccard distances < 0.45 were combined, with the combined term, keeping the name of the smallest GO term. We utilized the reduced set of GO terms to produce the heatmap.

3. Results and Discussion

3.1. Chemical Characterization of PSD Extracts: Parallel PAH Contamination, Divergent Qualitative Screen

Parallel PSD extracts allowed for exquisite PAH characterization using deuterated PRS and surrogate standards in the analytical extracts without fear of interference in the developmental toxicity assays. Undiluted extracts contained the chemicals concentrated in five HDPE strips over 30 days at each of the sampling locations. We report the nominal PAH concentrations in the 1% extract exposures. These PAH concentrations were approximately 1000 times higher than the concentration in the PHSS and were intended to identify differences in toxicity between mixtures, rather than mimic the environmental exposures. The sum PAH concentrations (Figure 1A) in each sample were 190 ± 13 and 310 ± 20 µM (±95% CIs) in RM 7W and RM 6.5W, respectively. The concentrations of each PAH above the limit of quantification are displayed in Figure 1B. Despite the 60% difference in magnitude, the PAH ratios between the two samples were nearly identical for FLA/PYR, RET/CHR, and PHE/ANT (Supplementary Table S3). Given the similar ratios, it is likely that the PAH contamination sources are similar, if not the same, at the two sites. Thus, even though the quantitative chemical analysis was limited to 33 PAHs, the composition of all PAHs in the two mixtures are likely very similar.

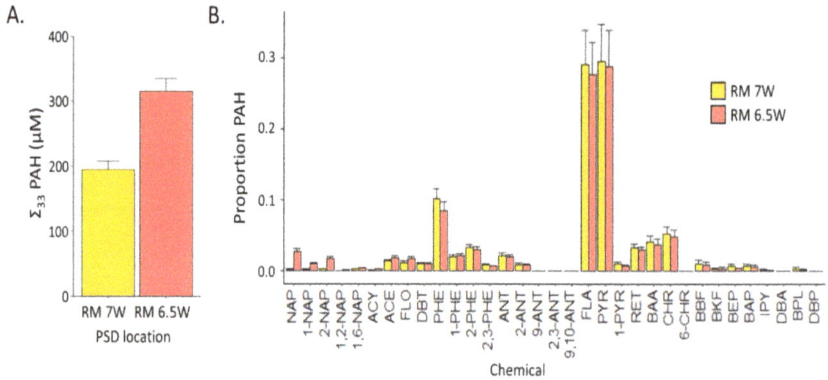

Figure 1. The results of the chemical analysis from the extracts of HDPE-PSDs deployed at RM 7W and RM 6.5 W during September 2009 and July 2010, respectively. The concentrations are reported at the expected levels in 1% extract exposures. (**A**) The sum uM of each PSD extract for the 33 PAHs included in the quantitative PAH method. (**B**) The proportion of each individual PAH measured in the 33 PAH method.

The 1201 chemical screen detected three chemicals in both extracts and five chemicals unique to only RM 7W (Table 1). Two DDT degradation products o,p'DDD and p,p'-DDE, along with hexachlorobenzene were detected in both samples. PCB65, PCB118, the synthetic musk tonalide, and two OPAHs benzofluorenone and BEZO. DDT degradants and hexachlorobenzene are both known PHSS contaminants, likely due to their use as pesticides [1]. More surprising is that despite having similar PAH profiles and higher concentrations of PAHs in RM 6.5W, there were more chemicals detected in RM 7W. While the PAH contaminant sources are similar, this is not necessarily true for other contaminants that may have contributed to the extract toxicity.

Table 1. Results of the 1201 chemical screen for RM 7W and RM 6.5W. Positive identification is indicated by an X.

Chemical	Detection	
	RM 7W	**RM 6.5W**
o,p'-DDD	X	X
p,p'-DDE	X	X
Hexachlorobenzene	X	X
PCB65	X	
PCB118X	X	
Tonalide	X	
Benzoflourenone	X	
benzanthrone	X	

3.2. Developmental Toxicity: RM 6.5W Was More Toxic Than RM 7W

Developmental toxicity assays demonstrated that RM 6.5W was more toxic than RM 7W and identified notochord malformations as the most sensitive developmental endpoint. Dechorionated zebrafish embryos were statically exposed to the vehicle control or to four dilutions of each PHSS extract and observed for negative outcomes at 24 and 120 hpf as a binary hit or no-hit response. As evident in Figure 2A, neither extract exerted developmental toxicity significantly different from the controls by 120 hpf at the two lowest concentrations, and both extracts exerted 100% incidence of malformation ("any effect except mortality") or mortality at the 1% exposure. In contrast, exposure to RM 6.5W at the 0.2% dilution caused malformations in 67.5% of the tested embryos, significantly more than RM 7W, which caused malformations in 26.5% of fish (Fisher's exact, $p < 0.05$). This data indicate that the developing zebrafish were more sensitive to RM 6.5W than RM 7W, and therefore, the nonpolar bioavailable fraction of contaminants at RM 6.5W was more toxic than the RM 7W site for the PSD deployment dates.

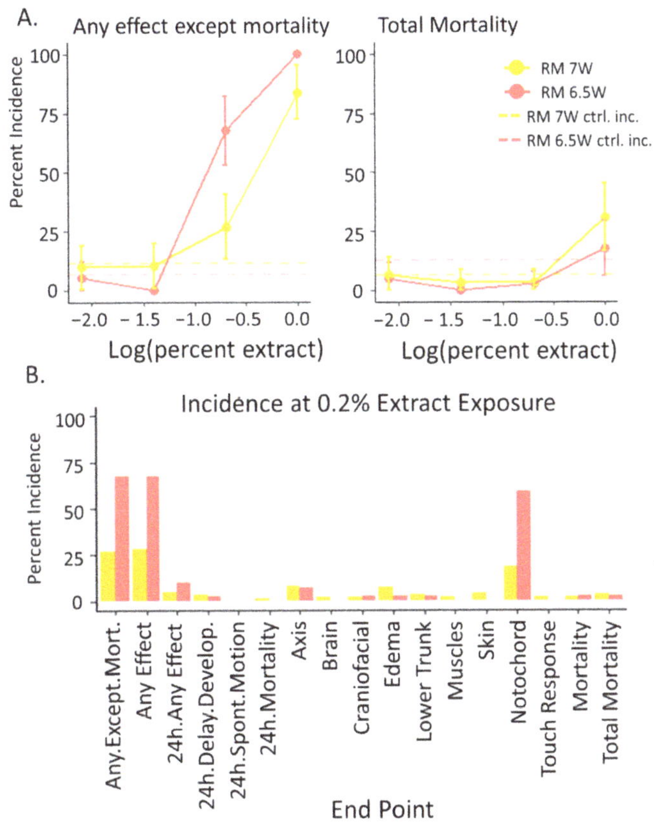

Figure 2. Concentration response data for the embryos exposed to the PSD extracts from RM 7W (September 2009) and RM 6.5W (July 2010) in the PHSS determined by the response of 40 developing zebrafish to each dose. (**A**) The percent incidence in the endpoints "any effect except mortality" and "mortality" in zebrafish by 120 hpf. Error bars indicate the 95% confidence intervals calculated utilizing the binomial distribution with n = 40. The hashed lines indicated the response levels in the vehicle controls associated with each mixture. (**B**) The observed percent incidence of every measured endpoint determined at the 0.2% extract exposure.

The increase in any effect except mortality from RM 6.5W compared to RM 7W and the controls was driven primarily by the wavy notochord malformation (Figure 2B), an unusual occurrence in developing zebrafish and not characteristic of PAH or AHR ligand exposure. Representative images for these phenotypes are included in the results in the study by Allan et al. [3]. In a developmental screen of 123 PAH and PAH-derivates, this malformation was not associated with any parent or alkyl PAHs and was only associated with four OPAHs, only one of which was an AHR ligand [38]. While two OPAHs were detected in RM 7W, neither elicited the wavy notochord response. The additional contaminants from RM 7W did not increase the toxicity of this mixture compared to RM 6.5W. Instead, the increased toxicity of RM 6.5W was consistent with its greater PAH concentration. Since PAHs alone did not explain the phenotypes observed in the developing embryos, it is likely that other toxicants in the mixture mirrored, to some degree, the PAH concentrations to drive toxicity.

3.3. *RNA-Seq: Direct Comparison of PHSS Extract Responses*

3.3.1. RM 6.5W Elicits a More Robust Gene Expression Response than RM 7W

RM 6.5W elicited a greater transcriptional response than RM 7W in embryos exposed from 6 to 48 hpf. We conducted RNA-Seq on three independent pools of 20 fish for RM 6.5W and RM 7W exposed at the 0.75% extract and an equal number of vehicle controls exposed to 1% DMSO. This extract dilution was chosen because it caused a high percentage of effect in larvae at 120 hpf without a significant effect at 48 hpf in both the extract treatments. These were exposed at equal extract dilutions so that the gene expression would reflect the relative bioactivity of each mixture.

Exposure to RM 7W resulted in 506 DEGs ($|logFC| > 1$ and $p_{adj} < 0.05$) while exposure to RM 6.5W resulted in 963 DEGs (Figure 3A). The directionality of the gene expression changes between the two samples was well-conserved. We compared the gene expression responses associated with exposure to each sample by plotting the differential expression in a heatmap, as shown in Figure 3C. There were almost no significant gene expression changes with different directionality between the two samples. Of the 1003 DEGs significant in either of the treatments, only 53 had different directional changes, and of these, only two were significant in both treatments. This was similar to the number of false positives we expected given $p_{adj} = 0.05$. The deeper colors in the RM 6.5W column illustrate the larger transcriptional responses to RM 6.5W.

To compare the magnitudes of gene expression, we plotted the log_2FC of genes with differential expression in either of the samples in Figure 3B. The expression patterns associated with the two samples were highly correlated. Fitting a linear model through the origin yielded a slope of 0.59 ± 0.02 (95% CI) with an $r^2 = 0.74$. This clearly indicated stronger gene expression changes resulting from the RM 6.5W exposure rather than the RM 7W exposure. The strong correlation in gene expression between the two mixtures was evidence that they exerted toxicity in a similar fashion. It is likely that the greater toxicity of RM 6.5W compared to RM 7W was caused by the same or similar components at higher concentrations affecting similar toxic pathways with more intensity.

The greater gene expression changes associated with RM 6.5W and the correlations in differential expression between the two exposures were consistent with those observed in the concentration–response evaluations. Namely, at the same dilution factors, the more toxic mixture evoked a larger transcriptional response. Used in this simple top–down approach, transcriptional responses in developing zebrafish exposed to PSD extracts clearly identified the more toxic mixture without having to exhaustively measure the mixture components. Up to this point, PAH concentrations have been associated with mixture toxicity, but have not been proven to drive it. Next, we used the DEGs to identify the potential negative outcomes to the developing embryos caused by these mixtures and their consistency with known PAH toxicity.

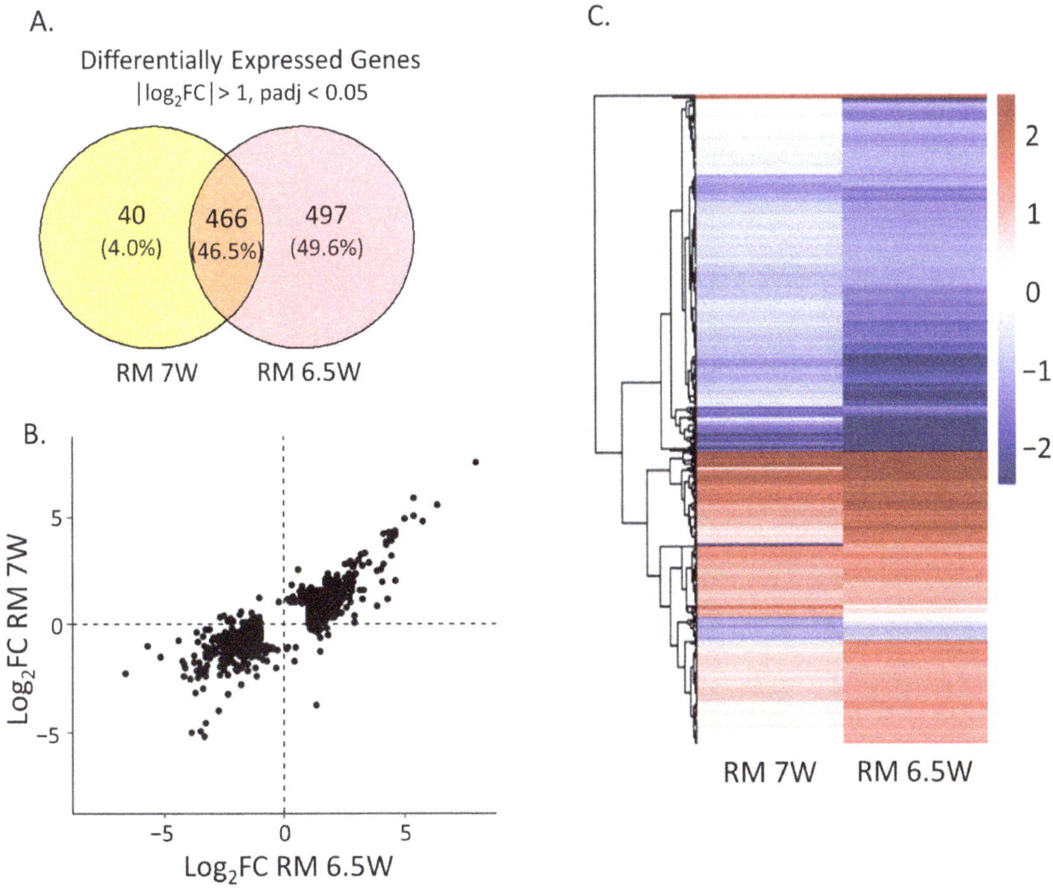

Figure 3. Differential expression in the embryos at 48 hpf after static exposure to the 0.75% extracts in embryo media from 6–48 hpf compared to the DMSO exposed control embryos. (**A**) The numbers of unique and shared differentially expressed genes (DEGs) meeting the threshold of $|\log_2 FC| > 1$, $p_{adj} > 0.05$ for each sample. (**B**) Log_2FC of gene expression comparing the exposure conditions for any gene differentially expressed under at least one of the conditions. (**C**) Heatmap displaying the Log_2FC of each sample, with genes clustered hierarchically by Euclidian distance.

3.3.2. Distinct Themes Underly the Transcriptional Response to PHSS Extract Exposures

We used a Gene Ontology network approach to identify concerted biological themes within the transcriptomic response to the mixtures. In this analysis, we decreased the log_2FC cutoff for DEGs to $|\log_2 FC| < 0.5$ and split the DEGs into three groups: DEGs specific to RM 7W treatment (RM 7W set), DEGs shared by both treatments (overlap set), and DEGs specific to RM 6.5W treatment (RM 6.5W set). We tested for overrepresentation in these DEG sets using terms from the Gene Ontology Database (GO) and visualized the results using the Cytoscape app Enrichment Map. The resulting network is displayed in Figure 4. Each node is a GO term colored by gene set, and edges denote the relative quantity of genes shared between them. The nodes were arranged into distinct clusters. We circled and numbered clusters with more than three GO terms and reviewed the terms to determine the themes included in the table in Figure 4. The full network file is available in the Supplementary Materials.

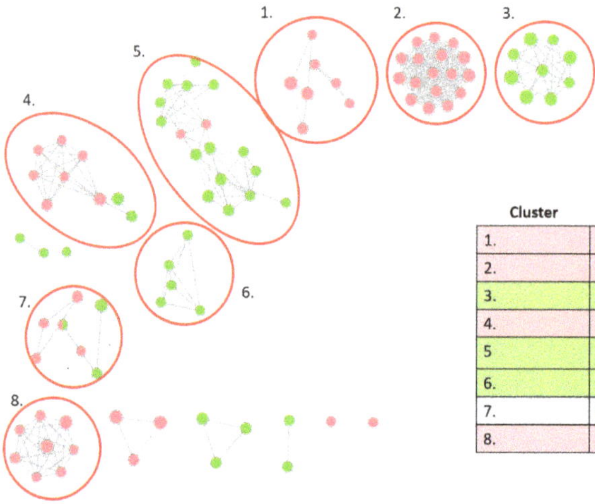

Overlap

RM 6.5W only

Cluster	Theme
1.	Anion transport
2.	Negative regulation of peptidase activity
3.	Visual perception and eye development
4.	tRNA aminoacylation
5	Xenobiotic response and detoxification
6.	Retinoid metabolism
7.	Hydrolase and glutathione metabolism
8.	Muscle fiber constituents

Figure 4. Gene Ontology (GO) network analysis. Each node is a significant GO term from one of three GO: Biological Processes (GO:BP), Cellular Components (GO:CC), or Molecular Functions (GO:MF). Ontologies were truncated to GO terms containing between 15 and 450 genes before computing significant adjuster enrichments on g.profiler2. Edges indicate the proportion of shared genes between terms, with an overlap coefficient ($C_{overlap} = n_{intersect}/n_{genes\ in\ smaller\ term}$) of at least 0.4. Node color indicates gene sets. Gene sets included DEGs meeting $|\log_2 FC| > 0.5$ and $p_{adj} > 0.05$, from one of three categories: DEGs specific to RM7 W, DEGs shared between the two treatments, and DEGs specific to RM 6.5W. There were no significant GO terms enriched in the gene set specific to RM 7W. The table indicates the manually determined themes for each cluster.

Surprisingly, the RM 7W set yielded no significantly enriched GO terms despite including 190 DEGs. The simplest explanation of this is that the DEGs unique to the RM 7W exposure did not represent any concerted pathway. Conversely, the overlap set resulted in 40 enriched GO terms. This vast disparity indicated that toxic processes caused by both PHSS extracts were significantly more important to RM 7W toxicity than any RM 7W-specific effect. Therefore, it did not appear that the additional contaminants detected in RM 7W affected its toxicity.

Another major trend was that the GO terms did not tend to be enriched in more than one DEG set. With the exception of a single GO term in cluster 7, every term was either from the overlap set or RM 6.5W set, but never both. This is likely because, by definition, the two DEG sets did not overlap. For a GO term to be enriched in both sets, it had to do so with entirely separate genes from both sets of DEGs.

The network distinguished between transcriptional responses unique to RM 6.5W exposure and those present in both PHSS extract exposures. GO terms diverged into distinct transcriptional clusters based on the shared genes in their terms. We divided clusters by which the DEG set tended to dominate among their GO terms. In this way clusters 3, 5, and 6 represent transcriptional themes elicited by exposure to both PHSS extracts, and clusters 1, 2, 4, and 8 are themes that emerged specifically out of RM 6.5W toxicity. These RM 6.5W emerging clusters could arise as downstream steps resulting from greater perturbation of earlier shared clusters, or they could represent completely distinct toxic processes to RM 6.5W. While not a certainty, the high correlation in transcriptional changes, even among genes that only met the DEG cutoff in RM 6.5W, makes the former explanation more likely. In order to decisively determine between them, we would need RNA measurements from embryos exposed to a phenotypically anchored concentration such as an EC80.

3.3.3. Xenobiotic Metabolism Glutathione Processes Are among the Expected GO Network Clusters

Clusters 5 and 7 contained GO terms and gene expression commonly associated with PAH exposure. Cluster 5 was bimodally distributed with terms relating to oxidant detoxification and xenobiotic metabolism. These two modes were connected by four GO terms, two from the RM 6.5W set and two from the overlap set, all dominated by genes for proteins containing hemes. Cluster 5 had the largest mean $|\log_2 FC|$ of any of the clusters, driven largely by classic AHR-responding genes such as cyp1a1, cyp1b1, ahrra, and foxq1a. The differential gene expression for these and other genes we call out throughout the study are included in Supplementary Table S4. RM 6.5W and RM 7W both contained PAHs, many of which were AHR ligands that elicited these responses in cell cultures and whole animal models including developing zebrafish [19,21,39]. Additionally, AHR is known to increase the transcription of nrf2, also upregulated in this cluster, which in turn regulates the response of many genes featured in the oxidant response GO terms [22].

Cluster 7, termed hydrolase and glutathione metabolism, also contained terms from both the overlap set and RM 6.5W set. The three terms in this cluster directly related to glutathione metabolism were enriched in the overlap set, while the four GO terms enriched in the RM 6.5W set were lyase related terms. The two groups were bridged by the terms "one-carbon metabolic process" and "cellular modified amino acid process", which each contained a few genes encoding lyases. The lyase-related GO terms did not provide easily interpretable insights because they were a broad class of enzymes and did not interrelate beyond having a similar catalytic mechanism. Glutathione related GO terms were similar to cluster 5. Glutathione is an important detoxifying agent [40]. As expected, all of the genes in this GO term, except for one, had increased transcription in response to both extract exposures.

3.3.4. Visual System Development and Muscle Fiber Related Genes Are Disrupted

Clusters 5 and 7 might be viewed as part of a reversible response by the developing zebrafish to the toxicants in the mixture. In evaluating the mixture toxicity, it would be more helpful to focus on clusters that are more likely to represent irreversible adverse outcomes. Clusters 3 and 8 likely correspond to irreversible toxic outcomes.

Cluster 3, visual perception and eye development, indicated significant visual system impairment. It was made entirely of GO terms from the overlap set, meaning that it was active in both PHSS extract exposures. Of the 60 genes in this cluster, 53 had decreased expression. Thirty-four were crystallin genes, which, when transcribed, form the proteins that make the lens of the eye [41]. Additionally, opsin 1, rhodopsin, and two other genes involved in photo transduction rom1b and arr3a had $\log_2$FC levels between −6.6 and −2.2 from RM 6.5W treatment, and −2.2 and −0.86 from RM 7W treatment, respectively. The widely decreased expression in crystallin genes and photo transduction genes indicated impaired eye development. This effect might have been mediated by Cyp1b1 expression induced by AHR. Cyp1B1 is naturally expressed in the eyes of zebrafish, mice, and humans, where it is known to play a role in development likely through the metabolism of a yet undetermined endogenous signaling molecule [42–45]. Despite this, the effects on the visual system are not classically implicated in AHR-mediated toxicity. While it is possible that the mixture of PAHs might have induced these effects through AHR, other constituents of the mixture could have acted through alternative means.

Cluster 8, themed muscle fiber constituents, also likely represented an irreversible adverse outcome and was interesting in light of the wavy notochord malformation induced so strongly by RM 6.5W exposure. It was the only cluster in the network that consisted mostly of GO terms related to Cellular Components, with a total of 25 genes enriched in six muscle cell constituent related terms. Cluster 8 also includes the GO terms "actin cytoskeleton" and "muscle cell development", which are broader in scope. Differential expression trended negatively, much like the visual perception cluster. For RM 6.5W exposure, the average $\log_2$FC of all DEGs in the muscle fiber cluster was −0.49 and 50 out of 60 DEGs had negatively impacted expression. The DEG with the largest fold change in this cluster was myl2b, an ortholog of human MLC2. It encodes a regulatory myosin light chain, which plays a role in potentiating muscle contraction and is essential for normal heart development in both zebrafish and mice [46,47]. Zebrafish with truncated mlc-2 or morpholino knocked-down mlc2 do not survive past seven days due to a lack of myofibril genesis [46]. With this in mind, the 5-fold reduction in mlc2 observed in embryos exposed to RM 6.5W very likely led to significant cardiac impairment. This cluster is particularly interesting in light of the wavy notochord phenotype caused by RM 6.5W exposure. Widespread disruption of structural components within these GO terms may have indicated diminished structural integrity within the embryos, resulting in a wavy notochord. PAHs are known cardiotoxins, with higher molecular weight PAHs typically acting through AHR signaling while lower molecular PAH toxicity is often independent of AHR; however, the widespread disruption of muscle fiber constituents has no precedence among PAHs [11,48,49]. The wavy notochord has previously been associated with a few OPAHS, two of which, 2-ethylanthraquinone and 3-hydroxyflouranthene, have parent PAHs measured in the PHSS extracts [38]. These data suggest that there are some PAH metabolites that can cause the phenotype. Exposure to a high concentration of a complex mixture of PAHs such as that of the PHSS extracts increases the likelihood that at least some of the metabolites could induce this phenotype. Conversely, exposure to a mixture of the ten most abundant PAHs in the PHSS extracts was not associated with this phenotype in previous studies [50]. The phenotype may also be caused by contaminants that went undetected by the 1201 chemical screen. In a screen of 1006 Phase 1 and Phase 2 Toxcast chemicals, the wavy notochord was only noted in exposure to 16 compounds, seven of which contained thiocarbamate functional groups [51]. In an earlier study, 0.8 μM exposure of metam sodium, a dithiocarbamate, resulted in the wavy notochord, and caused abnormal muscle physiology around the spinal cord [52]. A follow-up study demonstrated the wavy notochord in developmental zebrafish exposures in all nine dithiocarbamates tested [53]. This suggests that the gene expression changes in the muscle constituent cluster and wavy notochord could be driven by dithiocarbamate containing compounds. We recommend future chemical analysis conducted within the PHSS include targeted analysis for this class of compounds.

3.4. RNA-Seq: Comparison of Mixtures to Individual PAH and OPAH Exposure Responses

We next interpreted the gene expression of PHSS extract exposures through comparison to that of individual PAHs and OPAH. Ret, BkF, and BbF were selected because they are bioactive PAHs present in the mixtures, and OPAHs were included because they were detected in the RM 7W extract and are known to form during PAH degradation in the environment [54].

3.4.1. Principal Analysis Indicates AHR as Second Strongest Determinants of Variation

PCA analysis distinguished the exposures by chemical class and AHR activation status, and showed that the PHSS extract exposures were more similar to OPAH exposures than to PAH exposures (Figure 5A). We defined DEGS has having $|\log_2 FC| > 1$ and $p_{adj} < 0.05$ and used a PCA analysis to assess the similarity of treatments. Along PC1, the PAHs grouped together, followed by OPAHs, then RM 7W, and finally RM 6.5W. The OPAHS were more similar to RM 7W than to the PAHs, and RM 7W was much more similar to the OPAH exposures than to RM 6.5W. It is tempting to conclude that PC1 resolved by chemical class, with RM 7W grouping with OPAHs, but a parallel explanation was that PC1 divided by exposure toxicity. For instance, PAHs were exposed at the EC80 concentration, OPAHs at the EC100, and the PHSS exposures likely above the EC100. To differentiate between these two explanations, exposure concentrations would need to be normalized to the same EC before measuring gene expression. PC2 resolved the strong AHR activating exposures from the weak AHR activating exposures. PHEQ and BEZO are weak AHR transcription activators, as demonstrated by the cyp1a $\log_2 FC < 1.5$, resulting from both exposures. In contrast, all other exposures in this study had a cyp1a $\log_2 FCs$ greater than 5. This trend was echoed for other AHR reporters such as cyp1b1 and ahrra. In the same way, BEZO and PHEQ were much lower than the rest of the exposures along PC2. PC1 accounted for 53% of the variance between samples while PC2 accounted for 20 % of the variance in gene expression. The ability of PC2 to resolve strong from weak AHR activators indicated that AHR contributed to a significant portion of the gene expression response, but given the much greater strength of PC1 and the much larger difference between the two PHSS extracts in PC1, there were clearly other factors more responsible for driving the differences in transcription and toxicity among the PHSS extracts.

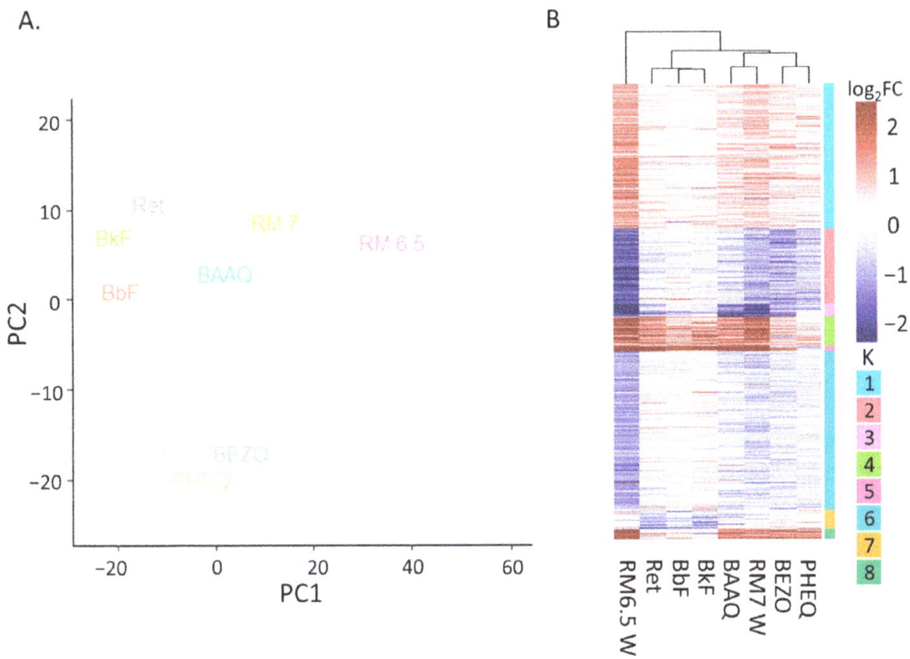

Figure 5. Comparison of the DEGs from the PHSS extract treatments and treatment with PAHs or OPAHs. (**A**) PCA analysis using DEGs of each treatment. PC1 accounts for 53% of variance, PC2 accounts for 20 % of variance. (**B**) A heatmap displaying $\log_2$FCs for each gene significant in at least one of the treatments ($|\log_2 FC| > 1$ and $p_{adj} <= 0.05$). Rows are grouped by k-means (k = 8) and columns are clustered by Euclidian distance.

3.4.2. Differential Expression Heatmap Contains AHR Transcription and OPAH Related Clusters Yet PHSS Extract Exposure Remains Distinct

Figure 5B is a heatmap for any gene that was differentially expressed in at least one treatment. The exposures were clustered hierarchically according to their differential expression, and by k-means. Again, RM 7W clustered with the OPAHs, the PAHs clustered together, and RM 6.5W had the most unique gene expression.

The differential expression from PHSS extract exposures followed the PAHs most closely in K2 and K3, but was largely distinct from these throughout the rest of the heatmap. K2 and K3 included many genes involved in xenobiotic metabolism and detoxification such as glutathione transferase ugt5a, P450s cyp1a1, cyp1a2, and cyp1a3, and other classic AHR response genes such as wfikkn1, ahrra, ahrrb, and the transcription factor foxq1a. In the remaining K groups, excluding K6, the average differential expression associated with PHSS exposures was higher in magnitude than that associated with individual PAH exposures. Similar to the relationship between RM 6.5W and RM 7W, the extra DEGs in the PHSS exposures might have resulted from differences in dosing. The zebrafish embryos were exposed to PAHs at the corresponding EC80 concentration, while the exposures to the PHSS extracts were likely at or above the EC100. Synergistic mixture effects among PAHs might also explain the differing expression. For instance, zebrafish co-exposed to the PAHs benzo(a)pyrene and fluoranthene, the former an AHR ligand, the latter a P450 inhibitor, experience a high level of cardiotoxicity not otherwise observed in exposure to either PAH alone [55]. We also cannot rule out other compounds within the mixture such as p-DDE and o-DDE, or compounds without good matches in the mass spectrum database as causative agents of these changes. Although DEGs induced by individual PAHs are present, the majority of transcriptional responses associated with exposure to the PHSS extracts was not accounted for by individual PAH exposure or single ligand–AHR activation.

The DEGs of both PHSS extracts were more closely related to DEGs from the OPAH exposures. K1 and K4–K8 displayed similar gene expression changes among the OPAH and PHSS extracts. In each of these groups, the DEGs were greater in number and magnitude for RM 6.5W exposure, followed by RM 7W, and finally the OPAHs. Given that OPAHs were only detected in RM 7W, it seemed peculiar that the OPAHs mirrored the PHSS extract expression better than the PAHs. It is unlikely that this occurred due to erroneous RNA-Seq results because they largely agree with other existing gene expression results for these compounds (Supplementary Table S5). Differential expression measured with qRT-PCR of cyp1a, cyp1b1, and akr1c1 in hepg2 cells exposed to BAAQ mirrored our RNA-Seq results [56]. Furthermore, a previous study by our group utilized qRT-PCR to examine the expression of common xenobiotic response genes with exposure to OPAH. A comparison to those results showed no disagreement among the DEGs determined in both techniques [57]. The apparent conundrum might arise from limitations in the non-exhaustive 1201 chemical screen. It is possible that some OPAHs were present in RM 6.5W but were not part of the screen. Likewise, some other undetected chemical or chemicals could have induced the OPAH-like gene expression. Alternatively, as suggested earlier, parent PAHs may be metabolized to OPAHs in sufficient quantities within the embryos to elicit similar effects to OPAH exposure.

3.4.3. There Is Significant Overlap between Individual Constituent and Mixture GO Terms but Muscle Fiber Related Genes Are without a Match

Finally, we hypothesized that GO term enrichment analysis could identify shared processes between the PHSS exposures and individual chemical processes. The analysis showed that GO terms resulting from both PHSS exposures could largely be found in individual OPAH and PAH exposures, while some RM 6.5W specific terms were still unaccounted for. Figure 6 summarizes the enrichment analysis results for the top 10 GO terms for each treatment excluding redundant terms.

There were 24 unique GO terms enriched in the DEGs from at least one of the PHSS extracts, 18 of which were also present in at least one PAH or OPAH exposure. With the exception of one GO term, all terms shared between the two PHSS extracts were also elicited by PAH or OPAH exposure. The most shared GO terms were "response to oxidative stress", "cellular response to xenobiotic stimulus", and "heme binding," all of which were terms from Cluster 5 in the network analysis. This indicates that many shared processes in the toxicity of the two PHSS extracts can be accounted for by shared or similar toxicants within the mixtures.

There were three visual perception related GO terms resulting from PHSS exposures, each of which was also associated with BEZO exposure. BEZO exposure resulted in 34 DEGs also present in the visual perception terms of the PHSS exposures. The linear model relating $\log_2$FC of these DEGs between BEZO and RM 6.5W was significant ($p \leq 0.05$) with $R^2 = 0.9$ and largely echoed by a model relating these DEGs between BEZO and RM 7W. Without filtering for significant DEGs in the BEZO exposure, there was still a significant correlation in the differential expression between the BEZO and the PHSS samples ($p \leq 0.05$). BEZO is a weak AHR transcriptional activator, and cyp1b1 is not a DEG associated with BEZO exposure, however, BEZO was able to induce a remarkably similar toxicity to the visual system. While not confirming a mechanism, this indicates that disruption to the visual system associated with PHSS exposure may be Cyp1B1 independent.

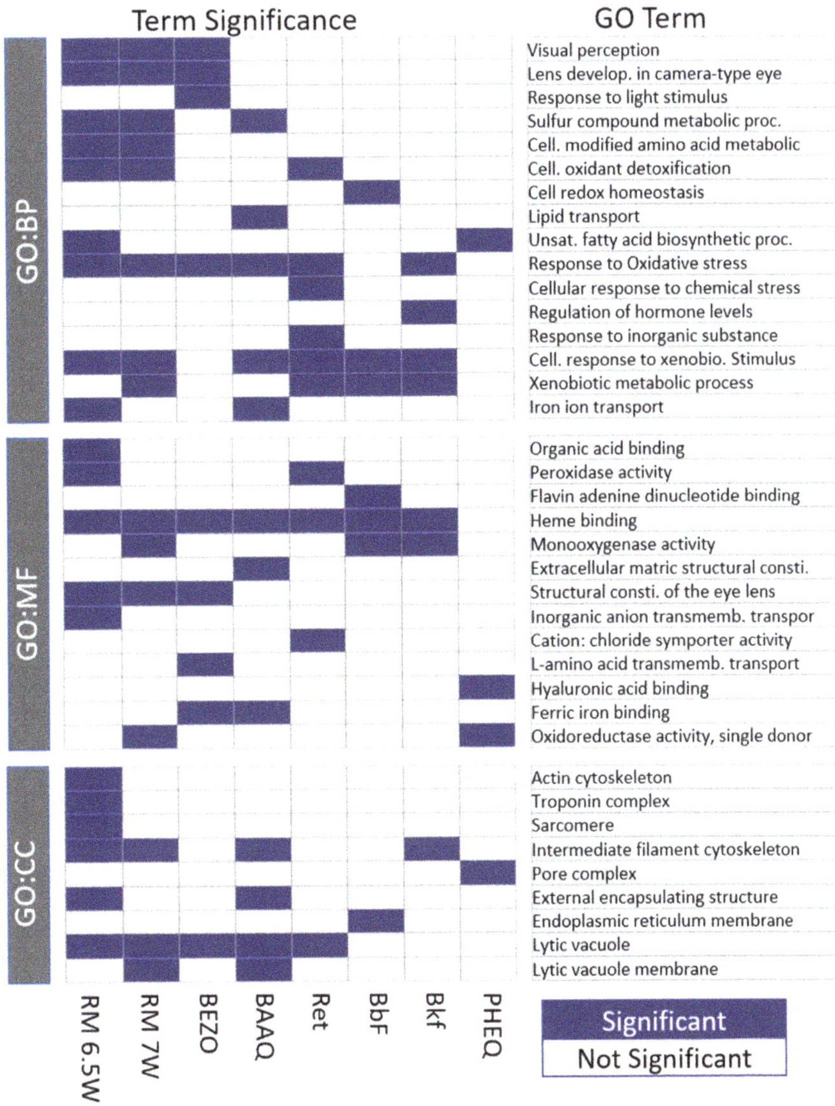

Figure 6. The significant GO terms from the reduced redundancy GO set. Enrichment was tested for the DEGs meeting the criteria of $|\log_2 FC| > 0.5$, $p_{adj} < 0.5$ for each exposure. Column order was chosen to simplify the comparison between the PHSS extracts and individual chemical exposures. Rows are split by the GO database then clustered by the Jaccard distance of shared genes.

While many GO terms enriched in DEGs from PHSS extract exposures were represented in the PAH or OPAH exposures, a few GO terms, most notably "actin cytoskeleton", "troponin complex" and "sarcomere" were unique to RM 6.5W exposure. These terms harkened to the muscle cell constituent cluster from the network analysis. Their lack of enrichment in the DEGs from the other exposures increased the likelihood that this particular toxicity arose from non-AHR ligand constituents of the mixture.

4. Conclusions

The trifecta of RNA-Seq, chemical analysis, and toxicity screening using PSD extracts and developing zebrafish proved to be a powerful approach to characterize hazards in whole environmental mixtures. Transcriptional responses provided a non-targeted method to identify perturbed biological processes underlying the gross malformations resulting from exposure to PHSS extracts. High correlation between differentially expressed genes in RM 7W and RM 6.5W indicated that similar xenobiotics affecting similar pathways drove the toxicity of both mixtures. Alone, the analytical chemistry and embryonic morphology assays detected adverse outcomes suggestive of PAH bioactivity, but the added dimension of transcriptomics uncovered not only PAH transcriptional signatures, but also perturbation of the visual and musculature systems uncharacteristic of canonical AHR ligand toxicity. Here, the differential gene expression of both PHSS mixtures was more similar to OPAHs than to PAHs, but also identified toxicities suggesting that mixture effects of other contaminants might be at play. Despite the suggestive gene expression data, true identification of the causal toxicants would likely require effects directed analysis, which was beyond the scope of this study.

The interpretation of our data was made more difficult because the exposures from which we measured the transcriptional responses were set at equal extract dilutions rather than to anchored to a phenotype. This resulted in the more toxic mixture exposure eliciting a larger transcriptional response, but came with an increased difficulty in specifying which transcriptional differences were truly unique between the two samples, and which differences were more likely to be due to the RM 6.5W exposure being "further up" the concentration–response curve. For the same reason, it also somewhat complicated our comparison to the differential expression resulting from individual compounds. To provide more certainty in discerning transcriptional differences, future studies would be better served by anchoring the mixture exposures to concentrations causing a predetermined level of phenotypic response such as the EC80. In typical circumstances where little is known about the mixture compositions, RNA-Seq could be used as a sort of fingerprint for the worst actors in the mix of chemicals. As more relevant comparative transcriptomic datasets between environmental mixtures and chemicals standards become available, such fingerprinting will become increasingly informative.

Supplementary Materials: The following supporting information can be downloaded at: https://www.mdpi.com/article/10.3390/toxics11030201/s1, Supplementary Figure S1: 1% PHSS extract pAH concentrations. Supplementary Table S1: PAH names and structures included in 33 PAH Quantitative Method, Supplementary Table S2: Summary of the exposure scenarios and techniques for RNA acquisition and sequencing; Supplementary Table S3: Diagnostic PAH ratios. Supplementary; Supplimentary Table S4: Differential Expresion of genes named in paper; Supplementary Table S5: Comparison of OPAH gene expression to results from other studies. Data S1: PAH concentrations in PHSS extracts. S2: Dose response for each endpoint, extract, and concentration. S3: Differential expression of each exposure scenario. GO term network: phss_mixtures_cytoscape_network.cys.

Author Contributions: Conceptualization, C.I.R., S.C.T., K.A.A. and R.L.T.; Methodology, C.I.R., L.G.T., K.A.A., K.M.W. and R.L.T.; Software, C.I.R. and L.G.T.; Validation, C.I.R. and L.G.T.; Formal analysis, C.I.R. and L.G.T.; Investigation, C.I.R.; Resources, K.A.A. and R.L.T.; Data curation, C.I.R. and K.M.W.; Writing—original draft preparation, C.I.R.; Writing—review and editing, C.I.R., L.G.T., S.C.T., K.M.W., K.A.A. and R.L.T.; Visualization, C.I.R. and K.M.W.; Supervision, S.C.T., K.M.W., K.A.A. and R.L.T.; Project administration, C.I.R.; Funding acquisition, K.M.W., K.A.A. and R.L.T. All authors have read and agreed to the published version of the manuscript.

Funding: This research was funded by the National Institutes of Health, P30 ES030287, P42 ES016465, and T32 ES007060.

Institutional Review Board Statement: The animal study protocol was approved by the Institutional Animal Care and Use Committee of Oregon State University, protocol 2020-0136, approved 9 December 2020.

Informed Consent Statement: Not applicable.

Data Availability Statement: The data presented in this study are available in the Supplementary Materials or upon request.

Acknowledgments: The authors would like to acknowledge the staff at the Sinnhuber Aquatic Research Laboratory who made this research possible. The Pacific Northwest National Laboratory is operated by Battelle for the U.S. Department of Energy under contract DE-AC05-76RL0 1830.

Conflicts of Interest: The authors declare no conflict of interest.

References

1. USEPA. *Portland Harbor Remedial Investigation Report, Final*; USEPA: Washington, DC, USA, 2016.
2. USEPA. *Revised Working Draft Sampling Plan for Pre-Remedial Design, Baseline and Long-Term Monitoring*; USEPA: Washington, DC, USA, 2017.
3. Allan, S.E.; Smith, B.W.; Tanguay, R.L.; Anderson, K.A. Bridging environmental mixtures and toxic effects. *Environ. Toxicol. Chem.* **2012**, *31*, 2877–2887. [CrossRef]
4. Hillwalker, W.E.; Allan, S.E.; Tanguay, R.L.; Anderson, K.A. Exploiting lipid-free tubing passive samplers and embryonic zebrafish to link site specific contaminant mixtures to biological responses. *Chemosphere* **2010**, *79*, 1–7. [CrossRef] [PubMed]
5. Lubin, A.; Otterstrom, J.; Hoade, Y.; Bjedov, I.; Stead, E.; Whelan, M.; Gestri, G.; Paran, Y.; Payne, E. A versatile, automated and high-throughput drug screening platform for zebrafish embryos. *Biol. Open* **2021**, *10*, bio058513. [CrossRef] [PubMed]
6. Adams, R.G.; Lohmann, R.; Fernandez, L.A.; MacFarlane, J.K. Polyethylene Devices: Passive Samplers for Measuring Dissolved Hydrophobic Organic Compounds in Aquatic Environments. *Environ. Sci. Technol.* **2007**, *41*, 1317–1323. [CrossRef] [PubMed]
7. Carls, M.G.; Holland, L.G.; Short, J.W.; Heintz, R.A.; Rice, S.D. Monitoring polynuclear aromatic hydrocarbons in aqueous environments with passive low-density polyethylene membrane devices. *Environ. Toxicol. Chem.* **2004**, *23*, 1416–1424. [CrossRef]
8. Minick, D.J.; Anderson, K.A. Diffusive flux of PAHs across sediment-water and water-air interfaces at urban superfund sites. *Environ. Toxicol. Chem.* **2017**, *36*, 2281–2289. [CrossRef] [PubMed]
9. Some non-heterocyclic polycyclic aromatic hydrocarbons and some related exposures. *IARC Monogr. Eval. Carcinog Risks Hum.* **2010**, *92*, 1–853.
10. Boffetta, P.; Jourenkova, N.; Gustavsson, P. Cancer risk from occupational and environmental exposure to polycyclic aromatic hydrocarbons. *Cancer Causes Control* **1997**, *8*, 444–472. [CrossRef]
11. Brette, F.; Shiels, H.A.; Galli, G.L.J.; Cros, C.; Incardona, J.P.; Scholz, N.L.; Block, B.A. A Novel Cardiotoxic Mechanism for a Pervasive Global Pollutant. *Sci. Rep.* **2017**, *7*, 41476. [CrossRef]
12. Das, L.; Patel, B.; Patri, M. Adolescence benzo[a]pyrene treatment induces learning and memory impairment and anxiolytic like behavioral response altering neuronal morphology of hippocampus in adult male Wistar rats. *Toxicol. Rep.* **2019**, *6*, 1104–1113. [CrossRef]
13. Incardona, J.P. Molecular Mechanisms of Crude Oil Developmental Toxicity in Fish. *Arch. Environ. Contam Toxicol.* **2017**, *73*, 19–32. [CrossRef] [PubMed]
14. McCallister, M.M.; Maguire, M.; Ramesh, A.; Aimin, Q.; Liu, S.; Khoshbouei, H.; Aschner, M.; Ebner, F.F.; Hood, D.B. Prenatal exposure to benzo(a)pyrene impairs later-life cortical neuronal function. *Neurotoxicology* **2008**, *29*, 846–854. [CrossRef] [PubMed]
15. Melendez-Colon, V.J.; Luch, A.; Seidel, A.; Baird, W.M. Formation of stable DNA adducts and apurinic sites upon metabolic activation of bay and fjord region polycyclic aromatic hydrocarbons in human cell cultures. *Chem. Res. Toxicol.* **2000**, *13*, 10–17. [CrossRef] [PubMed]
16. Novosad, J.; Fiala, Z.; Borska, L.; Krejsek, J. Immunosuppressive effect of polycyclic aromatic hydrocarbons by induction of apoptosis of pre-B lymphocytes of bone marrow. *Acta Med. (Hradec Kralove)* **2002**, *45*, 123–128. [CrossRef]
17. Denison, M.; Helferich, W. The Ah Receptor Signal Transduction Pathw ay. In *Toxicant-Receptor Interactions*; CRC Press: Boca Raton, FL, USA, 1998; pp. 15–46.
18. Miao, W.; Hu, L.; Scrivens, P.J.; Batist, G. Transcriptional regulation of NF-E2 p45-related factor (NRF2) expression by the aryl hydrocarbon receptor-xenobiotic response element signaling pathway: Direct cross-talk between phase I and II drug-metabolizing enzymes. *J. Biol. Chem.* **2005**, *280*, 20340–20348. [CrossRef] [PubMed]
19. Nebert, D.W.; Dalton, T.P.; Okey, A.B.; Gonzalez, F.J. Role of aryl hydrocarbon receptor-mediated induction of the CYP1 enzymes in environmental toxicity and cancer. *J. Biol. Chem.* **2004**, *279*, 23847–23850. [CrossRef] [PubMed]
20. Shankar, P.; Garcia, G.R.; La Du, J.K.; Sullivan, C.M.; Dunham, C.L.; Goodale, B.C.; Waters, K.M.; Stanisheuski, S.; Maier, C.S.; Thunga, P.; et al. The Ahr2-Dependent wfikkn1 Gene Influences Zebrafish Transcriptome, Proteome, and Behavior. *Toxicol. Sci.* **2022**, *187*, 325–344. [CrossRef]
21. Shankar, P.; Geier, M.C.; Truong, L.; McClure, R.S.; Pande, P.; Waters, K.M.; Tanguay, R.L. Coupling Genome-wide Transcriptomics and Developmental Toxicity Profiles in Zebrafish to Characterize Polycyclic Aromatic Hydrocarbon (PAH) Hazard. *Int. J. Mol. Sci.* **2019**, *20*, 2570. [CrossRef]
22. Yeager, R.L.; Reisman, S.A.; Aleksunes, L.M.; Klaassen, C.D. Introducing the "TCDD-inducible AhR-Nrf2 gene battery". *Toxicol. Sci.* **2009**, *111*, 238–246. [CrossRef]

23. Abdel-Shafy, H.I.; Mansour, M.S. A review on polycyclic aromatic hydrocarbons: Source, environmental impact, effect on human health and remediation. *Egypt. J. Pet.* **2016**, *25*, 107–123. [CrossRef]
24. Sower, G.J.; Anderson, K.A. Spatial and Temporal Variation of Freely Dissolved Polycyclic Aromatic Hydrocarbons in an Urban River Undergoing Superfund Remediation. *Environ. Sci. Technol.* **2008**, *42*, 9065–9071. [CrossRef]
25. Anderson, K.A.; Sethajintanin, D.; Sower, G.; Quarles, L. Field trial and modeling of uptake rates of in situ lipid-free polyethylene membrane passive sampler. *Environ. Sci. Technol.* **2008**, *42*, 4486–4493. [CrossRef] [PubMed]
26. Matzke, M.M.; Allan, S.E.; Anderson, K.A.; Waters, K.M. An approach for calculating a confidence interval from a single aquatic sample for monitoring hydrophobic organic contaminants. *Environ. Toxicol. Chem.* **2012**, *31*, 2888–2892. [CrossRef] [PubMed]
27. Garry Mallard, J.R. *Automated Mass Spectral Deconvolution and Identification System*; US National Institute of Standards and Technology: Gaithersburg, MD, USA, 1997; p. 57.
28. Goodale, B.C.; La Du, J.; Tilton, S.C.; Sullivan, C.M.; Bisson, W.H.; Waters, K.M.; Tanguay, R.L. Ligand-Specific Transcriptional Mechanisms Underlie Aryl Hydrocarbon Receptor-Mediated Developmental Toxicity of Oxygenated PAHs. *Toxicol. Sci.* **2015**, *147*, 397–411. [CrossRef]
29. Andrews, S. *FastQC: A Quality Control Tool for High Throughput Sequence Data, Version 0.11.9*; Babraham Bioinformatics: Cambridge, UK, 2015.
30. Dobin, A.; Davis, C.A.; Schlesinger, F.; Drenkow, J.; Zaleski, C.; Jha, S.; Batut, P.; Chaisson, M.; Gingeras, T.R. STAR: Ultrafast universal RNA-seq aligner. *Bioinformatics* **2013**, *29*, 15–21. [CrossRef] [PubMed]
31. Putri, G.H.; Anders, S.; Pyl, P.T.; Pimanda, J.E.; Zanini, F. Analysing high-throughput sequencing data in Python with HTSeq 2.0. *Bioinformatics* **2022**, *38*, 2943–2945. [CrossRef] [PubMed]
32. R Core Team. *R: A Language and Environemnt for Statistical Computing*; R Foundation for Statistical Computing: Vienna, Austria, 2021.
33. Love, M.I.; Huber, W.; Anders, S. Moderated estimation of fold change and dispersion for RNA-seq data with DESeq2. *Genome Biol.* **2014**, *15*, 550. [CrossRef]
34. Raudvere, U.; Kolberg, L.; Kuzmin, I.; Arak, T.; Adler, P.; Peterson, H.; Vilo, J. g:Profiler: A web server for functional enrichment analysis and conversions of gene lists (2019 update). *Nucleic Acids Res.* **2019**, *47*, W191–W198. [CrossRef]
35. Harris, M.A.; Clark, J.; Ireland, A.; Lomax, J.; Ashburner, M.; Foulger, R.; Eilbeck, K.; Lewis, S.; Marshall, B.; Mungall, C.; et al. The Gene Ontology (GO) database and informatics resource. *Nucleic Acids Res.* **2004**, *32*, D258–D261. [CrossRef]
36. Merico, D.; Isserlin, R.; Stueker, O.; Emili, A.; Bader, G.D. Enrichment map: A network-based method for gene-set enrichment visualization and interpretation. *PLoS ONE* **2010**, *5*, e13984. [CrossRef]
37. Shannon, P.; Markiel, A.; Ozier, O.; Baliga, N.S.; Wang, J.T.; Ramage, D.; Amin, N.; Schwikowski, B.; Ideker, T. Cytoscape: A software environment for integrated models of biomolecular interaction networks. *Genome Res.* **2003**, *13*, 2498–2504. [CrossRef]
38. Geier, M.C.; Chlebowski, A.C.; Truong, L.; Massey Simonich, S.L.; Anderson, K.A.; Tanguay, R.L. Comparative developmental toxicity of a comprehensive suite of polycyclic aromatic hydrocarbons. *Arch. Toxicol.* **2018**, *92*, 571–586. [CrossRef] [PubMed]
39. Chang, Y.; Siddens, L.K.; Heine, L.K.; Sampson, D.A.; Yu, Z.; Fischer, K.A.; Löhr, C.V.; Tilton, S.C. Comparative mechanisms of PAH toxicity by benzo[a]pyrene and dibenzo[def,p]chrysene in primary human bronchial epithelial cells cultured at air-liquid interface. *Toxicol. Appl. Pharmacol.* **2019**, *379*, 114644. [CrossRef]
40. Hayes, J.D.; Chanas, S.A.; Henderson, C.J.; McMahon, M.; Sun, C.; Moffat, G.J.; Wolf, C.R.; Yamamoto, M. The Nrf2 transcription factor contributes both to the basal expression of glutathione S-transferases in mouse liver and to their induction by the chemopreventive synthetic antioxidants, butylated hydroxyanisole and ethoxyquin. *Biochem. Soc. Trans.* **2000**, *28*, 33–41. [CrossRef] [PubMed]
41. Greiling, T.M.; Clark, J.I. The transparent lens and cornea in the mouse and zebra fish eye. *Semin. Cell Dev. Biol.* **2008**, *19*, 94–99. [CrossRef] [PubMed]
42. Bejjani, B.A.; Xu, L.; Armstrong, D.; Lupski, J.R.; Reneker, L.W. Expression patterns of cytochrome P4501B1 (Cyp1b1) in FVB/N mouse eyes. *Exp. Eye Res.* **2002**, *75*, 249–257. [CrossRef]
43. Doshi, M.; Marcus, C.; Bejjani, B.A.; Edward, D.P. Immunolocalization of CYP1B1 in normal, human, fetal and adult eyes. *Exp. Eye Res.* **2006**, *82*, 24–32. [CrossRef]
44. Williams, A.L.; Bohnsack, B.L. Neural crest derivatives in ocular development: Discerning the eye of the storm. *Birth Defects Res. C Embryo Today* **2015**, *105*, 87–95. [CrossRef]
45. Williams, A.L.; Eason, J.; Chawla, B.; Bohnsack, B.L. Cyp1b1 Regulates Ocular Fissure Closure Through a Retinoic Acid-Independent Pathway. *Investig. Ophthalmol. Vis. Sci.* **2017**, *58*, 1084–1097. [CrossRef]
46. Rottbauer, W.; Wessels, G.; Dahme, T.; Just, S.; Trano, N.; Hassel, D.; Burns, C.G.; Katus, H.A.; Fishman, M.C. Cardiac myosin light chain-2: A novel essential component of thick-myofilament assembly and contractility of the heart. *Circ. Res.* **2006**, *99*, 323–331. [CrossRef]
47. Sheikh, F.; Lyon, R.C.; Chen, J. Getting the skinny on thick filament regulation in cardiac muscle biology and disease. *Trends Cardiovasc. Med.* **2014**, *24*, 133–141. [CrossRef]
48. Korashy, H.M.; El-Kadi, A.O. The role of aryl hydrocarbon receptor in the pathogenesis of cardiovascular diseases. *Drug Metab. Rev.* **2006**, *38*, 411–450. [CrossRef] [PubMed]
49. Rigaud, C.; Eriksson, A.; Rokka, A.; Skaugen, M.; Lihavainen, J.; Keinänen, M.; Lehtivuori, H.; Vehniäinen, E.R. Retene, pyrene and phenanthrene cause distinct molecular-level changes in the cardiac tissue of rainbow trout (Oncorhynchus mykiss) larvae, part 2-Proteomics and metabolomics. *Sci. Total Environ.* **2020**, *746*, 141161. [CrossRef] [PubMed]

50. Geier, M.C.; James Minick, D.; Truong, L.; Tilton, S.; Pande, P.; Anderson, K.A.; Teeguardan, J.; Tanguay, R.L. Systematic developmental neurotoxicity assessment of a representative PAH Superfund mixture using zebrafish. *Toxicol. Appl. Pharmacol.* **2018**, *354*, 115–125. [CrossRef] [PubMed]
51. Truong, L.; Reif, D.M.; St Mary, L.; Geier, M.C.; Truong, H.D.; Tanguay, R.L. Multidimensional in vivo hazard assessment using zebrafish. *Toxicol. Sci.* **2014**, *137*, 212–233. [CrossRef]
52. Haendel, M.A.; Tilton, F.; Bailey, G.S.; Tanguay, R.L. Developmental toxicity of the dithiocarbamate pesticide sodium metam in zebrafish. *Toxicol. Sci.* **2004**, *81*, 390–400. [CrossRef] [PubMed]
53. Tilton, F.; La Du, J.K.; Vue, M.; Alzarban, N.; Tanguay, R.L. Dithiocarbamates have a common toxic effect on zebrafish body axis formation. *Toxicol. Appl. Pharmacol.* **2006**, *216*, 55–68. [CrossRef]
54. Lundstedt, S.; White, P.A.; Lemieux, C.L.; Lynes, K.D.; Lambert, I.B.; Oberg, L.; Haglund, P.; Tysklind, M. Sources, fate, and toxic hazards of oxygenated polycyclic aromatic hydrocarbons (PAHs) at PAH-contaminated sites. *Ambio* **2007**, *36*, 475–485. [CrossRef]
55. Wassenberg, D.M.; Di Giulio, R.T. Synergistic embryotoxicity of polycyclic aromatic hydrocarbon aryl hydrocarbon receptor agonists with cytochrome P4501A inhibitors in Fundulus heteroclitus. *Environ. Health Perspect.* **2004**, *112*, 1658–1664. [CrossRef]
56. Misaki, K.; Matsui, S.; Matsuda, T. Metabolic enzyme induction by HepG2 cells exposed to oxygenated and nonoxygenated polycyclic aromatic hydrocarbons. *Chem. Res. Toxicol.* **2007**, *20*, 277–283. [CrossRef]
57. Knecht, A.L.; Goodale, B.C.; Truong, L.; Simonich, M.T.; Swanson, A.J.; Matzke, M.M.; Anderson, K.A.; Waters, K.M.; Tanguay, R.L. Comparative developmental toxicity of environmentally relevant oxygenated PAHs. *Toxicol. Appl. Pharmacol.* **2013**, *271*, 266–275. [CrossRef] [PubMed]

toxics

MDPI

Article

Dioxin Congener Patterns in Breast Milk Samples from Areas Sprayed with Herbicide during the Vietnam War 40 Years after the War Ended

Muneko Nishijo [1,*], Hoa Thi Vu [1,2], Tai Pham-The [2], Thao Ngoc Pham [1,2], Nghi Ngoc Tran [3], Hideaki Nakagawa [4] and Hisao Nishijo [5]

1 Department of Public Health, Kanazawa Medical University, Uchinada 920-0293, Japan; vuhoa5593hvqy@gmail.com (H.T.V.); phamngocthaovmmu@gmail.com (T.N.P.)
2 Biomedical and Pharmaceutical Research Centre, Vietnam Military Medical University, Hanoi 12108, Vietnam; taithuy@kanazawa-med.ac.jp
3 Ministry of Health, Vietnam Government, Hanoi 10060, Vietnam; nghi_tranngoc@yahoo.com
4 Department of Social and Environmental Medicine, Kanazawa Medical University, Uchinada 920-0293, Japan; hnakagaw@kanazawa-med.ac.jp
5 System Emotional Science, University of Toyama, Sugitani, Toyama 930-0194, Japan; nishijo@med.u-toyama.ac.jp
* Correspondence: ni-koei@kanazawa-med.ac.jp; Tel.: +81-762188093

Abstract: Large amounts of herbicides containing polychlorinated dibenzo-p-dioxins and dibenzofurans (PCDD/Fs) were sprayed in South Vietnam during the Vietnam War. Levels of PCDD/Fs in the environment of South Vietnam remained high even 40 years later. A total of 861 breast milk samples (597 from three areas sprayed with herbicides, Quang Tri, Da Nang, and Bien Hoa, and 264 from three unsprayed areas in North Vietnam) were collected between 2007 and 2015 and the PCDD/F concentrations in the samples were determined. Levels of TEQ-PCDD/Fs and 17 PCDD/F congeners were higher in the sprayed area samples than the unsprayed area samples. We found particular PCDD/F congener patterns for different areas. High tetrachlorodibenzo-p-dioxin (TCDD) concentrations were found in Bien Hoa, high TCDD and 1,2,3,6,7,8-hexadibenzo-p-dioxin concentrations were found in Da Nang, and high 1,2,3,4,6,7,8-heptadibenzo-p-dioxin concentrations were found in Quan Tri. High 1,2,3,4,7,8-hexadibenzofuran and 1,2,3,4,6,7,8-heptadibenzofuran concentrations were also found in Da Nang and Quang Tri. However, breast feeding may have caused associations between the TCDD and polychlorinated dibenzofuran congener concentrations. Advanced statistical analysis will need to be performed in future to assess the characteristic PCDD/F congener profiles in breast milk samples from areas of Vietnam previously sprayed with herbicides.

Keywords: dioxin; congener profile; breast milk; herbicide; Vietnam

Citation: Nishijo, M.; Vu, H.T.; Pham-The, T.; Pham, T.N.; Tran, N.N.; Nakagawa, H.; Nishijo, H. Dioxin Congener Patterns in Breast Milk Samples from Areas Sprayed with Herbicide during the Vietnam War 40 Years after the War Ended. *Toxics* **2022**, *10*, 323. https://doi.org/10.3390/toxics10060323

Academic Editor: Paweł Cyplik

Received: 20 April 2022
Accepted: 10 June 2022
Published: 13 June 2022

Publisher's Note: MDPI stays neutral with regard to jurisdictional claims in published maps and institutional affiliations.

1. Introduction

Between 1961 and 1971, during operations by the United States (US) Armed Forces in the Vietnam War, >71 × 10^6 L of herbicides were sprayed in Vietnam south of the demilitarized zone (the 17th parallel) [1]. Agent Orange, which contains large quantities of 2,3,7,8-tetrachlorodibenzo-p-dioxin (TCDD), the most toxic polychlorinated dibenzo-p-dioxin and dibenzofuran (PCDD/F) congener, comprised approximately two-thirds of the herbicides that were sprayed [2]. Several decades after herbicide spraying ceased, PCDD/F, particularly TCDD, concentrations in environmental media and human tissues remain higher in areas near former US airbases in Da Nang and Bien Hoa than in unsprayed areas in Vietnam [3–5].

Historical material provided by the US Department of Defense to the Vietnam Government National Steering Committee on Overcoming the Post-war Consequences of Toxic Chemicals indicate that 150,000 m^3 of herbicide containing 365,000 ppt of PCDD/Fs were aerially sprayed near the Da Nang airbase between 1965 and 1971 and that 515,000 m^3 of herbicide containing 962,560 ppt of PCDD/Fs were aerially sprayed near the Bien Hoa airbase between 1965 and 1971. These former US airbases in Vietnam are called PCDD/F contamination hot spots [6]. Hatfield Consultants and the Vietnam Government found extremely high TCDD concentrations in soil and sediment near the Da Nang and Bien Hoa airbases in 2007 and 2010, nearly 40 years after herbicide spraying ended [7,8].

The PCDD/F congeners, including TCDD, are persistent in the environment and resistant to metabolism. PCDD/Fs are lipophilic, so can accumulate in fatty tissues in biota. PCDD/F concentrations in human tissues can increase with age because the biological half-lives of PCDD/F congeners are long (7–11 years for TCDD) [9–11]. Breast milk contains fat and can therefore contain PCDD/Fs at high concentrations. Milk samples are more readily collected than blood samples from humans. Therefore, milk samples are often collected to allow human exposure to PCDD/Fs to be monitored and to provide data for epidemiological studies of the toxic effects of PCDD/Fs in humans (e.g., cancer, childhood neurodevelopment, diabetes, and endocrine disruption).

We collected 241 breast milk samples from nursing mothers in the Da Nang cohort recruited in 2008–2009, one month after delivery and found that toxic equivalent (TEQ)-PCDD/Fs were four-times higher than that in breast milk from nursing mothers in unsprayed areas. However, TCDD contributed only ~10% of the TEQ-PCDD/Fs in the breast milk samples from the Da Nang cohort even though the TCDD contributions were higher in the breast milk samples from the unsprayed areas [4]. The mean TEQ-PCDD/Fs levels for 210 breast milk samples collected from another dioxin hot spot near Bien Hoa airbase in 2012 was three-times higher than the mean level for breast milk samples from Hanoi, an unsprayed area, but lower than the mean level for breast milk from Da Nang [5]. However, the TCDD concentrations were higher in breast milk from Bien Hoa than in breast milk from Da Nang. TCDD contributed >25% of the TEQ-PCDD/Fs in the samples from Bien Hoa [5]. These results suggest that dioxin contamination of breast milk has different characteristics in Bien Hoa and Da Nang. We also found higher concentrations of PCDD/F congeners other than TCDD in breast milk from Da Nang [4] and Bien Hoa [5] than in breast milk from unsprayed areas. However, we have not previously assessed differences between the concentrations of PCDD/F congeners other than TCDD in samples from Da Nang and Bien Hoa.

In 2002–2003, we determined the concentrations of PCDD/F congeners in breast milk samples from mothers living in Quang Tri, which was heavily sprayed with herbicides during the Vietnam War but is located far from former US airbases. We found higher concentrations of high-chlorinated PCDD/F congeners such as hexachlorodibenzo-p-dioxins (HexaCDD) and heptachlorodibenzo-p-dioxin (HeptaCDD) in the samples from Quang Tri [12] than were found in breast milk samples collected in the 1970s and 1980s from Da Nang and Dong Nai (where the Bien Hoa airbase is located) by Schecter et al. [13] and Dwernychuk et al. [14]. These findings suggest that the PCDD/F congener patterns in the breast milk samples collected soon after Vietnam War might be different from the patterns in the breast milk samples collected almost 40 years after the war ended.

Therefore, we investigated PCDD/F congener patterns in breast milk samples from dioxin contamination hot spots nearby former US military airbases (Da Nang and Bien Hoa), a sprayed area only in the past time (Quang Tri), and unsprayed areas to clarify the characteristics of dioxin contamination in areas of Vietnam that have been sprayed with herbicides 40 years after the Vietnam War ended.

2. Materials and Methods

2.1. Study Areas and Subjects

Samples from the dioxin contamination hot spots around Da Nang and Bien Hoa airbases were analyzed. Samples from 227 mothers living in Thanh Khe and Son Tra districts in Da Nang city and <10 km from the Da Nang airbase who participated in a birth cohort study in 2008 and 2009 [15] were analyzed. Samples from 283 mothers living in 10 communes around Bien Hoa airbase (another hot spot) in Bien Hoa city and who gave birth at Dong Nai General Hospital were recruited to birth cohort studies in 2012 and 2015 [16] were analyzed. The locations of these hot spots are shown on a map of Vietnam in Figure 1.

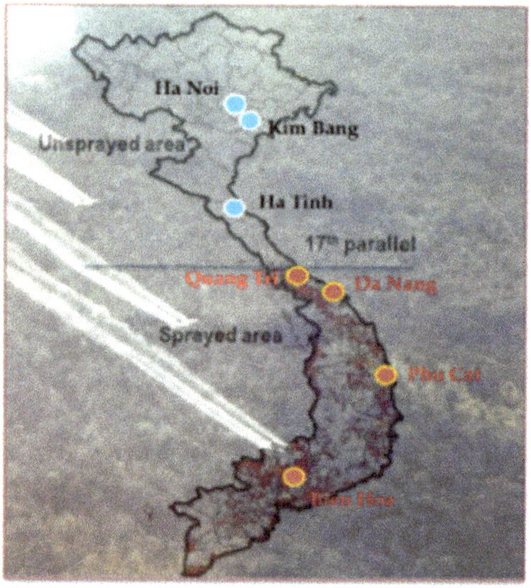

Figure 1. A map of Vietnam.

Samples from 87 mothers living in a sprayed area in Cam Chinh commune in Cam Lo district, Quang Tri province, who had been recruited to surveys in 2002 and 2003 [12,17] were analyzed. Quang Tri province is ~30 km south of the 17th parallel, the demilitarized zone (see Figure 1), and was heavily sprayed during the Vietnam War.

Samples from the unsprayed areas marked blue in Figure 1 were also analyzed. Samples from 62 mothers in the Cam Phuc commune in Cam Xuyen district, Ha Tinh province, ~280 km north of the 17th parallel enrolled in a survey in 2003 were analyzed. Samples from 75 mothers in Kim Bang district, Ha Nam province, ~50 km from the center of Hanoi city in northern Vietnam, enrolled in a survey in 2008 [15,18,19] were analyzed. Samples from 127 mothers in Ha Dong district in Hanoi city enrolled in a birth cohort study in 2014 [16] were also analyzed. None of the unsprayed areas had been affected by industrial pollution and were treated as areas with only background dioxin contamination.

In total, 861 breast milk samples were analyzed. A total of 597 of the samples were from mothers in three dioxin contaminated areas, and the other 264 samples were from mothers in three areas with background levels of dioxins.

Written informed consent was obtained from all mothers in both areas according to a process reviewed and approved by the Health Department of Da Nang City and Dong Nai Province in Vietnam, Hanoi Medical University, and Vietnam Military Medical University. The institutional ethics board at Kanazawa Medical University (Japan) approved the design of this study (No. ES-187).

2.2. Determining Dioxin Concentrations in Breast Milk

We collected ~20 mL of breast milk from each mother at her home with assistance from community health station medical staff 1–6 months after the mother had given birth. Each milk sample was frozen and then transferred packed with dry ice to Kanazawa Medical University in Japan. The samples were then stored at −30 °C until the concentrations of 2,3,7,8-chlorinated PCDD/F congeners in the samples were determined.

Milk fat was extracted from each sample and the milk fat content of the sample was determined gravimetrically, then 2,3,7,8-chlorinated ^{13}C-labeled PCDD/F internal standards (DF-LCS-A40; Wellington Laboratories, Guelph, Canada) were added. The milk fat was then purified and fractionated using previously published methods [20], then the extract was analyzed using an HP-6980 gas chromatograph (Agilent Technologies, Santa Clara, CA, USA) equipped with an MStation-JMS700 high-resolution mass spectrometer (JEOL, Tokyo, Japan) to determine the concentrations of 17 PCDD and PCDF congeners. The methods used to determine the PCDD/F concentrations in the milk samples were described in previous publications [4,5,21]. For quality control, a certificated reference material of natural milk powder (CRM 607, Community Bureau of reference-BCR, European Commission, Belgium) diluted was analyzed after dilution in distilled water. Recovery rates of ^{13}C-labeled internal standards for every individual PCDD/F congener were within 50–115%. Limit of detection (LOD) was defined as a signal-to-noise (S/N) ratio of peak height of chromatogram = 3. LODs for each congener of our measurement are shown in Table 1. The values of the concentrations of congeners below LODs were set to half of the detection limits for statistical analysis.

Table 1. Adjusted mean concentrations of PCDD/F congeners in breast milk from mothers living in Quang Tri, Da Nang, and Bien Hoa, where herbicides were sprayed during the Vietnam War, and from mothers living in unsprayed areas in North Vietnam.

	LOD	Unsprayed areas $n = 264$				Quang Tri $n = 87$					Da Nang $n = 227$					Bien Hoa $n = 283$				
	(ppt)	Med	GM	GSD	Adj M	Med	GM	GSD	Adj M		Med	GM	GSD	Adj M		Med	GM	GSD	Adj M	
PCDD congeners (pg/g lipid)																				
2,3,7,8-TetraCDD	0.004	0.6	0.6	1.9	0.6	0.8	0.8	2.0	0.8	*	1.4	1.4	2.3	1.4	*	2.0	2.3	2.2	2.1	*
1,2,3,7,8-PentaCDD	0.011	1.3	1.2	1.7	1.3	2.3	2.3	1.9	2.3	*	4.2	4.1	1.6	4.1	*	2.9	2.9	1.6	2.7	*
1,2,3,4,7,8-HexaCDD	0.004	0.6	0.6	1.7	0.6	1.7	1.4	2.4	1.4	*	2.3	2.2	1.6	2.2	*	1.3	1.4	1.6	1.3	*
1,2,3,6,7,8-HexaCDD	0.010	1.4	1.3	1.6	1.3	6.7	6.2	1.8	6.2	*	8.4	8.0	1.7	8.1	*	4.5	4.4	1.8	4.3	*
1,2,3,7,8,9-HexaCDD	0.008	0.6	0.6	1.8	0.6	1.8	1.6	2.2	1.6	*	2.6	2.6	1.7	2.6	*	2.0	2.1	2.0	2.1	*
1,2,3,4,6,7,8-HeptaCDD	0.008	2.2	2.1	1.7	2.1	14.1	13.3	1.8	13.2	*	11.8	11.8	1.6	11.9	*	8.5	9.2	1.9	9.0	*
OctaCDD	0.008	14.7	13.1	2.2	13.4	45.2	44.1	1.8	44	*	64.2	66.7	1.6	67	*	55.2	61.7	1.9	59.8	*
PCDF congeners (pg/g lipid)																				
2,3,7,8-TetraCDF	0.005	0.8	0.8	1.6	0.8	0.5	0.5	1.7	0.5	*	0.5	0.5	2.0	0.5	*	0.6	0.6	1.7	0.6	*
1,2,3,7,8-PentaCDF	0.006	0.6	0.5	1.8	0.5	0.6	0.6	2.1	0.6		1.2	1.2	1.9	1.2	*	0.6	0.6	1.7	0.6	
2,3,4,7,8-PentaCDF	0.010	2.8	2.7	1.5	2.8	4.4	4.3	1.9	4.4	*	7.3	7.0	1.6	7.1	*	3.5	3.6	1.6	3.5	*
1,2,3,4,7,8-HexaCDF	0.008	1.7	1.6	1.8	1.7	14.4	12.6	2.1	12.7	*	17.8	16.9	1.8	17.2	*	5.7	5.8	1.6	5.6	*
1,2,3,6,7,8-HexaCDF	0.003	1.4	1.4	1.6	1.4	8.3	7.4	2.1	7.4	*	11.0	10.3	1.8	10.4	*	3.3	3.4	1.6	3.3	*
1,2,3,7,8,9-HexaCDF	0.004	0.3	0.3	2.1	0.3	1.0	1.0	2.3	1.0	*	0.2	0.2	2.3	0.3		0.4	0.5	1.8	0.5	*
2,3,4,6,7,8-HexaCDF	0.006	0.5	0.3	2.8	0.3	0.3	0.3	2.9	0.3		1.3	1.3	1.7	1.3	*	0.7	0.7	1.7	0.7	*
1,2,3,4,6,7,8-HeptaCDF	0.010	1.1	1.1	1.8	1.1	10.8	10.4	2.2	10.4	*	11.9	11.9	1.8	12.0	*	3.2	3.3	1.8	3.3	*
1,2,3,4,7,8,9-HeptaCDF	0.004	0.2	0.2	2.3	0.2	1.6	1.4	2.8	1.4	*	1.3	1.2	2.2	1.2	*	0.6	0.7	1.9	0.6	*
OctaCDF	0.013	0.5	0.5	3.6	0.5	0.2	0.2	3.3	0.2		0.5	0.6	2.4	0.6	*	1.6	1.6	2.1	1.6	*
TEQs (pg/g lipid)																				
TEQ-PCDD/Fs		3.7	3.7	1.5	3.8	9.7	9.1	1.8	9.1	*	12.8	12.5	1.6	12.7	*	8.5	9.0	1.7	8.6	*

LOD: limit of detection, n: number of subjects, Med: median, GM: geometric mean, GSD: geometric standard deviation, Adj M: mean adjusted for age and parity (1: primiparae, 2: multiparae); CDD: chlorodibenzo-p-dioxin, CDF: chlorodibenzofuran, TEQ: toxic equivalent, *: $p < 0.05$ at Bonferroni correction for multiple comparisons compared with unsprayed areas.

The TEQ value for each PCDD/F congener in a sample was calculated by multiplying the concentration of the congener by the toxic equivalence factor for the congener published by the World Health Organization in 2005 [22], and then the TEQ-PCDD/Fs level for the sample was calculated by adding together the TEQ values for all of the congeners.

2.3. Statistical Analysis

Statistical analyses were performed using SPSS version 22.0 software (IBM, Armonk, NY, USA).

The levels of 17 PCDD/F congeners and the TEQ-PCDD/F in the breast milk samples were logarithmically (base 10) transformed to make the distributions match normal distributions more closely, and then the levels in the samples from Quang Tri, Da Nang, Bien Hoa, and the unsprayed areas were compared using general linear models after adjusting the data for maternal age (in years) and parity (primiparae and multiparae).

The concentrations of each PCDD/F congener in the samples from the three dioxin contaminated areas were stratified into low, middle, and high concentrations using the 95th percentile concentration for the samples from the unsprayed areas as the cut-off for the low and middle concentrations. The cut-off for the middle and high concentrations was the 75th percentile concentration for the Da Nang samples. The odds ratios (ORs) for high concentrations of four PCDD/F congeners in the samples from Da Nang and Bien Hoa were compared with the ORs for Quang Tri using a binary logistic regression model after adjusting for maternal age and parity (primiparae and multiparae).

Associations between the TCDD concentrations and the concentrations of four PCDD/F congeners were identified by a linear regression model for the different areas after adjusting for maternal age and parity for all mothers or adjusting for maternal age for primiparae or multiparae mothers.

For all tests, $p < 0.05$ was taken to indicate statistical significance.

3. Results

The mean concentrations of 17 PCDD/F congeners and TEQ-PCDD/F levels in the breast milk samples from mothers living in Quang Tri (an area in which herbicide was sprayed during the Vietnam War), the dioxin hot spots in Da Nang and Bien Hoa, and the unsprayed areas were compared after adjusting for age and parity, and the results are shown in Table 1. The TEQ-PCDD/F levels were significantly higher in the Quang Tri, Da Nang, and Bien Hoa samples than the unsprayed area samples. The concentrations of all of the PCDD/F congeners were significantly higher in the Bien Hoa samples than the unsprayed area samples. The concentrations of most of the PCDD/F congeners were significantly higher in the Da Nang and Quang Tri samples than the unsprayed area samples. However, the 1,2,3,7,8-pentachlorodibenzofuran (PentaCDF) and 2,3,4,6,7,8-hexachlorodibenzofuran (HexaCDF) concentrations were not significantly higher in the Quang Tri samples than the unsprayed area samples and the 1,2,3,7,8,9-HexaCDF concentrations were not significantly higher in the Da Nang samples than the unsprayed area samples.

The median TEQ-PCDD/F levels were more than two- or three-times higher in the Quang Tri, Da Nang, and Bien Hoa samples than in the unsprayed area samples. The median TCDD concentrations were three-times higher in the Da Nang and Bien Hoa samples than the unsprayed area samples, but the median TCDD concentration was only 1.3-times higher in the Quang Tri samples than the unsprayed area samples. The median concentrations of the other PCDD congeners were higher in the samples from the three exposed areas than the unsprayed areas. In particular, the median 1,2,3,6,7,8-HexaCDD and 1,2,3,4,6,7,8-HeptaCDD concentrations were higher in the Quang Tri and Da Nang samples than the Bien Hoa samples, as shown in Table 1. The median 1,2,3,4,7,8-HexaCDF, 1,2,3,6,7,8-HexaCDF, and 1,2,3,4,6,7,8-HeptaCDF concentrations were higher in the samples from the three exposed areas than the unsprayed areas, and the concentrations were two-times higher in the Quang Tri and Da Nang samples than the Bien Hoa samples, as shown in Table 1.

After assessing these median concentrations, we selected five congeners (TCDD, 1,2,3,6,7,8-HexaCDD, 1,2,3,4,6,7,8-HeptaCDD, 1,2,3,4,7,8-HexaCDF, and 1,2,3,4,6,7,8-HeptaCDF) and calculated the percentages of the samples from primiparae and multiparae mothers living in Quang Tri, Da Nang, Bien Hoa, and unsprayed areas that contained these congeners at low, middle, and high concentrations. The results are shown in Figures 2–4. The TCDD concentrations were high (≥2.3 pg/(g lipid)) in ~15%, ~25%, ~40%, and none of the samples from primipara mothers living in Quang Tri, Da Nang, Bien Hoa, and the unsprayed areas, respectively, as shown in Figure 2A. High TCDD concentrations were found in similar percentages of samples from multipara and primipara mothers living in Da Nang and Bien Hoa but fewer samples from multipara mothers (<5%) than primipara mothers living in Quang Tri, as shown in Figure 2B.

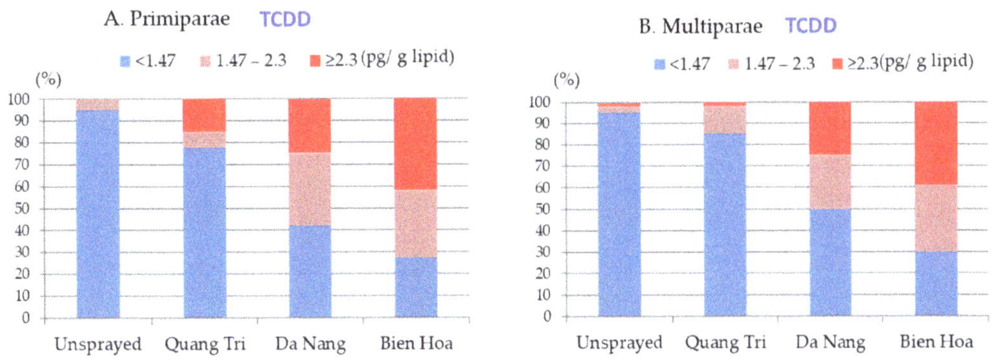

Figure 2. Percentages of samples from primiparae (**A**) and multiparae (**B**) mothers living in Quang Tri, Da Nang, Bien Hoa, and unsprayed areas containing TCDD at low, middle, and high concentrations.

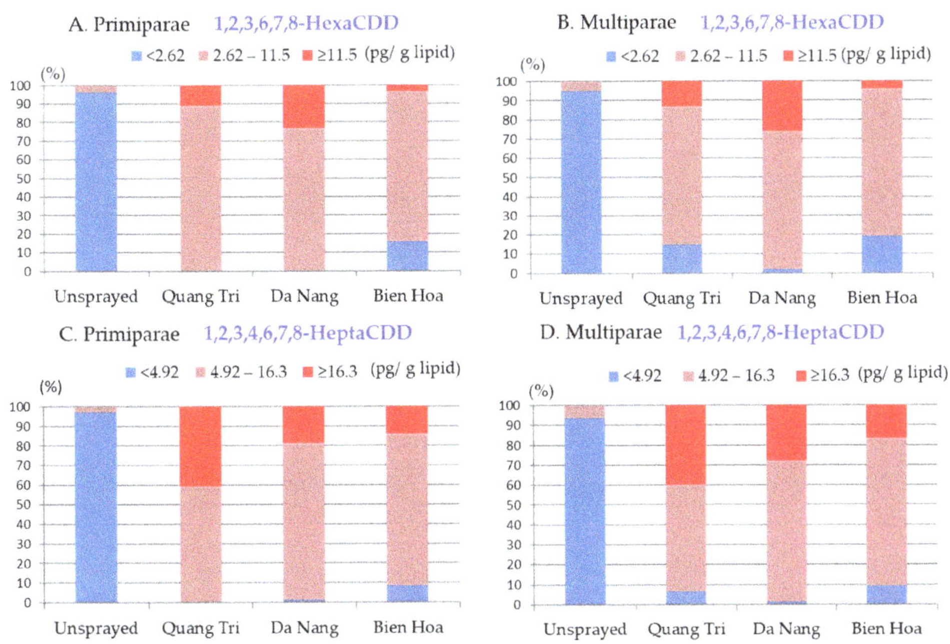

Figure 3. Percentages of samples from (**A**,**C**) primiparae and (**B**,**D**) multiparae living in Quang Tri, Da Nang, Bien Hoa, and unsprayed areas containing 1,2,3,6,7,8-HexaCDD (**A**,**B**) and 1,2,3,4,6,7,8-HeptaCDD (**C**,**D**) at low, middle, and high concentrations.

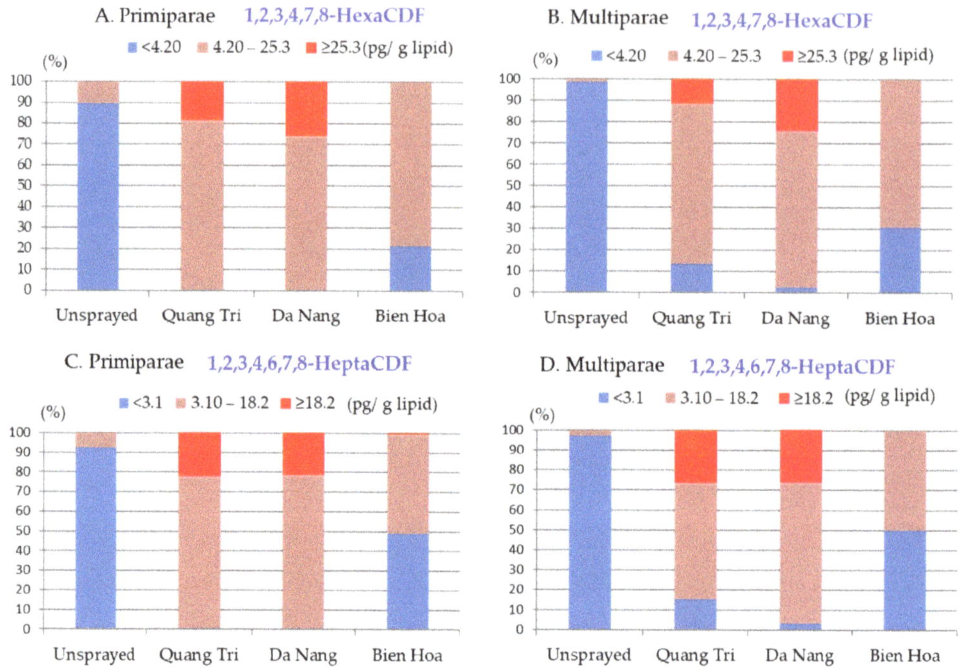

Figure 4. Percentages of samples from (**A**,**C**) primiparae and (**B**,**D**) multiparae living in Quang Tri, Da Nang, Bien Hoa, and unsprayed areas containing levels of 1,2,3,4,7,8-HexaCDF (**A**,**B**) and 1,2,3,4,6,7,8-HeptaCDF (**C**,**D**) at low, middle, and high concentrations.

Areas with statistically different percentages of samples containing high TCDD concentrations were identified using a binary logistic regression model after adjusting for maternal age and primipara and multipara mothers, and the results are shown in Table 2. For primipara mothers, the OR for a high TCDD concentration was significantly higher for Bien Hoa (OR 3.7) than the other areas, indicating that the risks of the TCDD concentration being high were similar for Da Nang and Quan Tri but significantly higher for Bien Hoa than Quang Tri. For multipara mothers, the ORs were high for Da Nang and Bien Hoa (ORs 19.5 and 36.6, respectively) relative to Quang Tri, indicating that the risks of the TCDD concentration being high were higher for Da Nang and Bien Hoa than Quan Tri. For all mothers, the OR for a high TCDD concentration after adjusting for age and parity was almost twice as high for Bien Hoa (OR 10.0) as for Da Nang (OR 5.6), indicating that the risk of the TCDD concentration being high was higher for Bien Hoa than the other herbicide-sprayed areas.

Table 2. Adjusted odds ratios (ORs) for high TCDD concentrations in breast milk from mothers living in Da Nang and Bien Hoa relative to those in breast milk from mothers living in Quang Tri.

	Quang Tri			Da Nang				Bien Hoa			
	n	%	OR	*n*	%	OR		*n*	%	OR	
Primiparae	4	14.8	1.0	17	24.6	2.1		47	41.6	3.7	*
Multiparae	1	1.7	1.0	39	24.7	19.5	**	66	38.8	36.6	***
All	5	5.7	1.0	56	24.7	5.6	***	113	39.9	10.0	***

n: number of subjects, OR: odds ratio, *: $p < 0.05$, **: $p < 0.01$, ***: $p < 0.001$ after adjusting for age or age and parity.

The 1,2,3,6,7,8-HexaCDD concentrations were high (≥11.5 pg/g lipid) in ~10%, ~25%, and <5% of the samples from primipara mothers living in Quang Tri, Da Nang, and

Bien Hoa, respectively, and middle 1,2,3,6,7,8-HexaCDD concentrations (2.6–11.5 pg/g lipid) were found in 85% of the samples from primipara mothers living in Bien Hoa, as shown in Figure 3A. The 1,2,3,6,7,8-HexaCDD concentrations were similar for samples from multipara and primipara mothers, although the percentages of samples with low concentrations were slightly higher for samples from multiparae than primiparae living in Quang Tri and Bien Hoa, as shown in Figure 3B. The 1,2,3,4,6,7,8-HeptaCDD concentrations were high (≥16.3 pg/g lipid) in 40%, 20–25%, and 15% of the samples from primipara and multipara mothers living in Quang Tri, Da Nang, and Bien Hoa, respectively, as shown in Figure 3C,D.

The ORs for high 1,2,3,6,7,8-HexaCDD concentrations in the Da Nang, and Bien Hoa samples relative to the Quang Tri samples are shown in Table 3. The ORs were similar for samples from primipara and multipara mothers living in Da Nang (2.6 and 2.2, respectively) and Bien Hoa (0.25 and 0.25, respectively), indicating that the risks were higher in Da Nang and lower in Bien Hoa than in Quang Tri. For both primipara and multipara mothers, the risks of high 1,2,3,6,7,8-HexaCDD concentrations were significantly higher in Da Nang and significantly lower in Bien Hoa than in Quang Tri.

Table 3. Adjusted odds ratios (ORs) for high 1,2,3,6,7,8-HexaCDD and 1,2,3,4,6,7,8-HeptaCDD concentrations in breast milk from mothers living in Da Nang and Bien Hoa relative to those in breast milk from mothers living in Quang Tri.

	Quang Tri			Da Nang				Bien Hoa			
	n	%	OR	n	%	OR		n	%	OR	
					1,2,3,6,7,8-HexaCDD						
Primiparae	3	11.1	1.0	16	23.2	2.6		4	3.5	0.25	
Multiparae	8	13.3	1.0	41	25.9	2.2		7	4.1	0.25	*
All	11	12.6	1.0	57	25.1	2.4	*	11	3.9	0.25	**
					1,2,3,4,6,7,8-HeptaCDD						
Primiparae	11	40.7	1.0	13	18.8	0.37		16	14.2	0.20	***
Multiparae	24	40.0	1.0	44	27.8	0.57		28	16.5	0.27	***
All	35	40.2	1.0	57	25.1	0.50	*	44	15.5	0.24	***

n: number of subjects, OR: odds ratio, *: $p < 0.05$, **: $p < 0.01$, ***: $p < 0.001$ after adjusting for age or age and parity.

The ORs for high 1,2,3,4,6,7,8-HeptaCDD concentrations in the Da Nang and Bien Hoa samples were low relative to the ORs for the Quan Tri samples (OR 1.0) for both primipara and multipara mothers, but the ORs were significantly lower only for the Bien Hoa samples. However, for both parity groups, the ORs for high 1,2,3,4,6,7,8-HeptaCDD concentrations were significantly lower for both the Da Nang and Bien Hoa samples than the Quang Tri samples, indicating that the risks of high 1,2,3,4,6,7,8-HeptaCDD concentrations were lower in Da Nang and Bien Hoa than Quang Tri, as shown in Table 3.

The 1,2,3,4,7,8-HexaCDF concentrations were high (≥25.3 pg/g lipid) in 10–20% of the Quang Tri samples and 25% of the Da Nang samples from primipara mothers (Figure 4A) and multipara mothers (Figure 4B), but none of the Bien Hoa samples had high concentrations. The 1,2,3,4,6,7,8-HeptaCDF concentrations were high (≥18.2 pg/g lipid) in similar percentages of the Quang Tri and Da Nang samples from primipara mothers (20%; Figure 4C) and multipara mothers (25%; Figure 4D). Very few of the Bien Hoa samples exhibited high 1,2,3,4,6,7,8-HeptaCDF concentrations, and around half of the samples from primipara and multipara mothers had middle concentrations, as shown in Figure 4C,D).

No samples from Bien Hoa contained high 1,2,3,4,7,8-HexaCDF and 1,2,3,4,6,7,8-HeptaCDF concentrations, so ORs were calculated only for Da Nang relative to Quang Tri (data not shown). However, the ORs for high 1,2,3,4,7,8-HexaCDF and 1,2,3,4,6,7,8-HeptaCDF concentrations were not significantly different for Da Nang (0.98 and 1.03, respectively) than Quang Tri, indicating similar risks of exposure to high 1,2,3,4,7,8-HexaCDF and 1,2,3,4,6,7,8-HeptaCDF concentrations in Da Nang and Quang Tri.

Associations between the TCDD concentrations and the concentrations of other PCDD/F congeners found at high concentrations in the sprayed areas (1,2,3,6,7,8-HexaCDD, 1,2,3,4,6,7,8-HeptaCDD, 1,2,3,4,7,8-HxCDF, and 1,2,3,4,6,7,8-HeptaCDF) were assessed after adjusting for maternal age and parity. Standardized β values for each area are shown in Table 4. Significant associations were found between the TCDD concentrations and the 1,2,3,6,7,8-HexaCDD, 1,2,3,4,6,7,8-HeptaCDD, 1,2,3,4,7,8-HxCDF, and 1,2,3,4,6,7,8-HeptaCDF concentrations for all of the areas, including the unsprayed areas. However, the highest β values for 1,2,3,6,7,8-HexaCDD, 1,2,3,4,6,7,8-HeptaCDD, and 1,2,3,4,7,8-HxCDF (i.e., not 1,2,3,4,6,7,8-HeptaCDF) were for Quang Tri, followed by Da Nang and Bien Hoa. In particular, the β values for all of the congeners were lower for Bien Hoa than the unsprayed areas and other sprayed areas. The β values for 1,2,3,4,6,7,8-HeptaCDF were lower for all of the sprayed areas than the unsprayed areas, indicating that the 1,2,3,4,6,7,8-HeptaCDF concentration was more weakly associated with the TCDD concentration in Quang Tri, Da Nang, and Bien Hoa than the areas at background concentrations.

Table 4. Adjusted associations between the TCDD concentrations and concentrations of other PCDD/F congeners that were found at high concentrations in the herbicide sprayed areas.

	Unsprayed Areas (n = 264)		**Dioxin Exposed Areas**					
			Quang Tri (n = 87)		**Da Nang** (n = 227)		**Bien Hoa** (n = 283)	
	β	***p*-Value**	**β**	***p*-Value**	**β**	***p*-Value**	**β**	***p*-Value**
1,2,3,6,7,8-HexaCDD	0.476	***	0.559	***	0.518	***	0.438	***
1,2,3,4,6,7,8-HeptaCDD	0.351	***	0.501	***	0.385	***	0.285	***
1,2,3,4,7,8-HexaCDF	0.332	***	0.429	***	0.392	***	0.258	***
1,2,3,4,6,7,8-HeptaCDF	0.332	***	0.229	*	0.271	***	0.205	**

n: number of subjects, β: standardized beta; CDD: chlorodibenzo-p-dioxin; CDF: chlorodibenzofuran; *: $p < 0.05$, **: $p < 0.01$, ***: $p < 0.001$ after adjusting for age and parity (1: primiparae, 2: multiparae).

The PCDD/F concentrations were lower in the samples from multiparae than primiparae, so associations between the TCDD concentrations and 1,2,3,6,7,8-HexaCDD, 1,2,3,4,6,7,8-HeptaCDD, 1,2,3,4,7,8-HxCDF, and 1,2,3,4,6,7,8-HeptaCDF concentrations in each area were assessed after adjusting for maternal age for primipara and multipara mothers. The results are shown in Table 5. The β values for all of the congeners for unsprayed areas were significant and were higher for primiparae than multiparae. However, the β value for 1,2,3,4,6,7,8-HeptaCDF for primipara mothers in Quang Tri was not significant and was lower than the β value for multipara mothers. The β values for 1,2,3,6,7,8-HexaCDD in Da Nang and 1,2,3,4,6,7,8-HeptaCDD in Bien Hoa were significant but lower for primiparae than multiparae. The β values for 1,2,3,4,7,8-HxCDF and 1,2,3,4,6,7,8-HeptaCDF for primipara mothers in Da Nang and Bien Hoa were not significant but the β values for multipara mothers were significant. These results indicated that TCDD concentrations and 1,2,3,6,7,8-HexaCDD, 1,2,3,4,6,7,8-HeptaCDD, 1,2,3,4,7,8-HxCDF, and 1,2,3,4,6,7,8-HeptaCDF concentrations may be associated in different ways for primipara and multipara mothers.

Differences in the associations between the TCDD concentrations and the 1,2,3,6,7,8-HexaCDD, 1,2,3,4,6,7,8-HeptaCDD, 1,2,3,4,7,8-HxCDF, and 1,2,3,4,6,7,8-HeptaCDF concentrations for primipara and multipara mothers were investigated by plotting the 1,2,3,4,6,7,8-HeptaCDD and 1,2,3,4,6,7,8-HeptaCDF concentrations against the TCDD concentrations in the samples from the sprayed and unsprayed areas. The plots are shown in Figures 5 and 6. In each scatter plot, data for the samples from the primipara and multipara mothers are indicated by purple circles and green circles, respectively. The β values for the relationships between the TCDD concentrations and 1,2,3,4,6,7,8-HeptaCDD and 1,2,3,4,6,7,8-HeptaCDF concentrations according to parity are shown in each plot.

Table 5. Age-adjusted associations between TCDD concentrations and concentrations of four other PCDD/F congeners according to maternal parity.

	Primiparae								Multiparae							
	Unsprayed Areas		Dioxin Exposed Areas						Unsprayed Areas		Dioxin Exposed Areas					
			Quang Tri		Da Nang		Bien Hoa				Quang Tri		Da Nang		Bien Hoa	
	n = 105		n = 27		n = 69		n = 113		n = 159		n = 60		n = 158		n = 170	
	β	p-Value	β	p-Value	β	p-Value	β	p-Value	β	p-Value	β	p-Value	β	p-Value	β	p-Value
1,2,3,6,7,8-HexaCDD	0.586	***	0.590	**	0.303	**	0.448	***	0.415	***	0.579	***	0.574	***	0.437	***
1,2,3,4,6,7,8-HeptaCDD	0.361	***	0.477	***	0.393	**	0.223	*	0.334	***	0.501	***	0.383	***	0.320	***
1,2,3,4,7,8-HexaCDF	0.401	***	0.478	***	0.214		0.170		0.303	***	0.435	**	0.439	***	0.307	***
1,2,3,4,6,7,8-HeptaCDF	0.384	***	0.094		0.139		0.090		0.311	***	0.261	*	0.308	***	0.267	**

β: standardized regression coefficient, n: number of subjects; CDD: chlorodibenzo-p-dioxin; CDF: chlorodibenzofuran; *: $p < 0.05$, **: $p < 0.01$, ***: $p < 0.001$ after adjusting for maternal age.

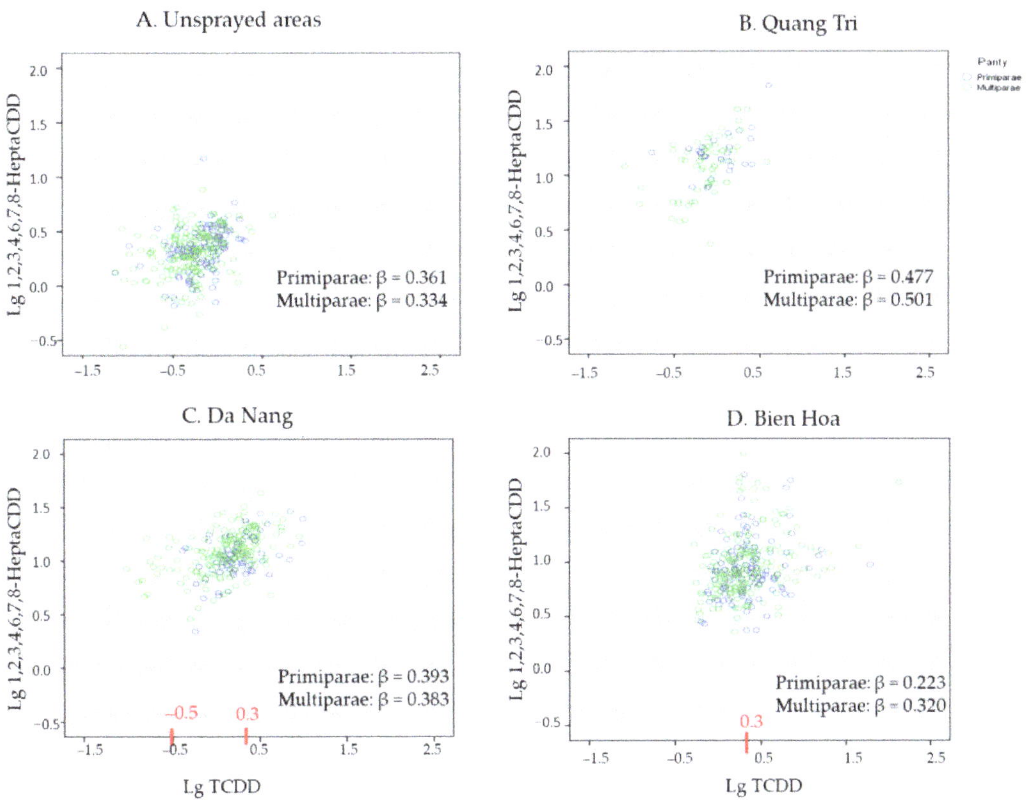

Figure 5. Relationships between the 1,2,3,4,6,7,8-HeptaCDD and TCDD concentrations (shown as base-10 logarithmically transformed values) in the samples from (**A**) unsprayed areas, (**B**) Quang Tri, (**C**) Da Nang, and (**D**) Bien Hoa. Purple circles denote samples from primipara mothers and green circles denote samples from multipara mothers.

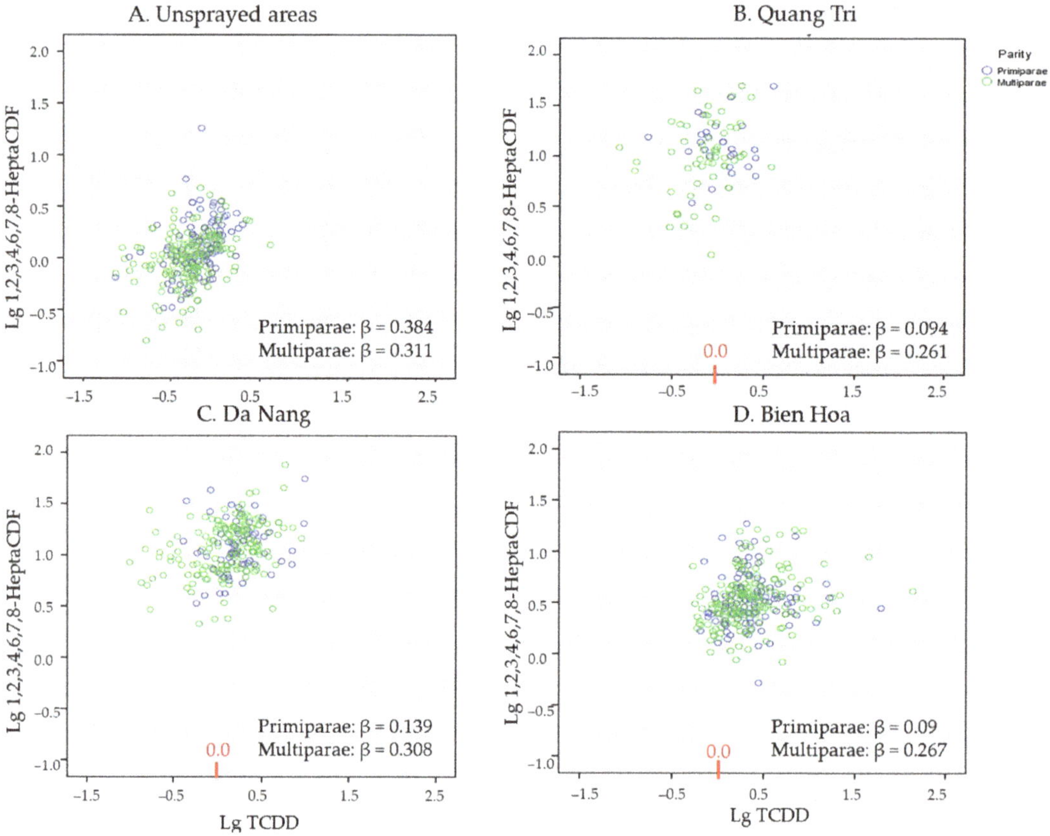

Figure 6. Relationships between the 1,2,3,4,6,7,8-HeptaCDF and TCDD concentrations (shown as base-10 logarithmically transformed values) in the samples from (**A**) the unsprayed areas, (**B**) Quang Tri, (**C**) Da Nang, and (**D**) Bien Hoa. Purple circles denote samples from primipara mothers and green circles denote samples from multipara mothers.

The 1,2,3,4,6,7,8-HeptaCDD concentration increased as the TCDD concentration increased for both primipara and multipara mothers in Quan Tri as shown in Figure 5B and Da Nang as shown in Figure 5C. However, the 1,2,3,4,6,7,8-HeptaCDD concentrations increased less in samples with high TCDD concentrations (>2.0 pg/g lipid) (0.3 log units) from primipara and multipara mothers from Da Nang than in samples with lower TCDD concentrations, as shown in Figure 5C. This made the association weaker for the samples with TCDD concentrations <0.3 pg/g lipid (−0.5 log units) from multipara mothers because the 1,2,3,4,6,7,8-HeptaCDD concentrations were relatively high. As shown in Figure 5D, some samples with TCDD concentrations >2.0 pg/g lipid (0.3 log units) from Bien Hoa had extremely high 1,2,3,4,6,7,8-HeptaCDD concentrations, meaning the TCDD and 1,2,3,4,6,7,8-HeptaCDD concentrations were weakly associated, particularly for the samples from primipara mothers.

No significant association was found between the TCDD and 1,2,3,4,6,7,8-HeptaCDF concentrations in the samples from primipara mothers from Quang Tri (β value 0.094; Figure 6B), Da Nang (β value 0.139; Figure 6C), or Bien Hoa (β value 0.09; Figure 6D). However, the samples with TCDD concentrations <1.0 pg/g lipid (log value 0.0) from multipara mothers had relatively low 1,2,3,4,6,7,8-HeptaCDF concentrations, resulting in weak but significant correlations between the TCDD and 1,2,3,4,6,7,8-HeptaCDF concentrations for all of the herbicide-exposed areas (see Figure 6B–D).

The 1,2,3,6,7,8-HexaCDD, 1,2,3,4,6,7,8-HeptaCDD, 1,2,3,4,7,8-HxCDF, and 1,2,3,4,6,7,8-HeptaCDF concentrations in the samples from the herbicide-exposed areas increased as the TCDD concentrations increased, but the strengths of the associations were different for the PCDDs and PCDFs and for samples from primipara and multipara mothers. The 1,2,3,6,7,8-HexaCDD and 1,2,3,4,6,7,8-HeptaCDD concentrations were strongly (more strongly for 1,2,3,6,7,8-HexaCDD than 1,2,3,4,6,7,8-HeptaCDD) associated with the TCDD concentration for the samples from all of the areas and both parity categories, although the association between the TCDD and 1,2,3,4,6,7,8-HeptaCDD concentrations in the samples from primipara mothers from Bien Hoa included samples with TCDD concentrations >2.0 pg/g lipid. No associations were found between the 1,2,3,4,7,8-HxCDF and 1,2,3,4,6,7,8-HeptaCDF concentrations and the TCDD concentrations in the samples from primipara mothers in Da Nang and Bien Hoa. The associations between the TCDD concentration and the 1,2,3,6,7,8-HexaCDD, 1,2,3,4,6,7,8-HeptaCDD, and 1,2,3,4,7,8-HxCDF concentrations were stronger in the samples from Quang Tri than the other areas, including the unsprayed areas, for both parity categories, but the association between the TCDD concentration and the 1,2,3,4,6,7,8-HeptaCDF was not, indicating that the congener profiles for the Quang Tri and Da Nang samples were not the same.

4. Discussion

4.1. Dioxin Concentrations in Breast Milk from Vietnamese Mothers

The results provide a broad view of the characteristics of PCDD/F congeners in milk from nursing mothers in areas with different levels of dioxin contamination in Vietnam. The TEQ-PCDD/F levels and the concentrations of almost all of the PCDD/F congeners, including TCDD, were significantly higher in the samples from the herbicide-exposed areas (Quang Tri, Da Nang, and Bien Hoa) in South Vietnam than in the samples from the unsprayed areas in North Vietnam. High concentrations of TCDD, 1,2,3,6,7,8-HexaCDD, 1,2,3,4,6,7,8-HeptaCDD, 1,2,3,4,7,8-HexaCDF, and 1,2,3,4,6,7,8-HeptaCDF were specific to herbicide-exposed areas, but characteristic congeners were different for each exposed area.

A high TCDD concentration was characteristic for Bien Hoa for primipara mothers and for Bien Hoa and Da Nang for multipara mothers. A high 1,2,3,6,7,8-HexaCDD concentration was particular for Da Nang but not for Bien Hoa for both parity categories. A high 1,2,3,4,6,7,8-HeptaCDD concentration was particular for Quang Tri. High 1,2,3,4,7,8-HexaCDF and 1,2,3,4,6,7,8-HeptaCDF concentrations were found only in samples from Da Nang and Quang Tri, but not in samples from Bien Hoa. These results indicated that Bien Hoa and Da Nang may be affected by PCDD/Fs in Agent Orange used at air bases but that herbicides other than Agent Orange containing PCDD/Fs with different congener profiles may have been used at Da Nang and at Quang Tri (in the middle of Vietnam).

According to the US Department of Defense, 52,700 containers of Agent Orange, 29,000 containers of Agent White, and 5000 containers of Agent Blue were prepared at Da Nang airbase between 1964 and 1972 and 98,000 containers of Agent Orange (almost twice as much as at Da Nang airbase), 45,000 containers of Agent White, and 16,000 containers of Agent Blue were used at Bien Hoa airbase between 1966 and 1972 [23]. This suggests that TCDD would have been the dominant PCDD/F at Bien Hoa. Other herbicides such as Agent White contained less TCDD and more other congeners and were sprayed in the middle of Vietnam by the US Air Force from Da Nang airbase.

There were some differences between the congener profiles in the samples from Quang Tri and Da Nang, and 1,2,3,4,6,7,8-HeptaCDD and 1,2,3,4,6,7,8-HeptaCDD concentrations were higher and more associated with the TCDD concentration for the Quang Tri samples than the Da Nang samples. These results suggested that PCDD/F exposure may be different in Quang Tri and Da Nang. In Quang Tri (which did not have a nearby airbase), PCDD/F contamination was probably caused by herbicide spraying during the Vietnam War. This would have caused TCDD concentrations to be lower than at PCDD/F contamination hot spots.

In the present study, it was found that strength of associations with increasing TCDD was different between primiparae and multiparae. Two PCDD congeners showed strong associations with TCDD in all areas and in both parity categories, except association between TCDD and 1,2,3,4,6,7,8-HeptaCDD in primipara mothers in Bien Hoa. For two PCDF congeners, however, associations with TCDD were found only in multipara mothers in Da Nang and Bien Hoa. For 1,2,3,4,6,7,8-HeptaCDF, association with TCDD was also observed in multipara mothers in Quan Tri. Since significant associations with TCDD were observed for all these congeners both in primipara and multipara mothers in unsprayed areas, decreased levels of dioxins in maternal body, particularly PCDF congeners with short biological half-life, through breast feeding might be related with significant associations only in multipara mothers.

The strengths of the associations between the PCDD/F congener concentrations and TCDD concentration were different for samples from primipara and multipara mothers. The 1,2,3,6,7,8-HexaCDD and 1,2,3,4,6,7,8-HeptaCDD concentrations were strongly associated with the TCDD concentrations in the samples from most of the areas and for both parity categories, but no association was found between the TCDD and 1,2,3,4,6,7,8-HeptaCDD concentrations in the samples from primipara mothers from Bien Hoa. Associations between the 1,2,3,4,7,8-HxCDF and 1,2,3,4,6,7,8-HeptaCDF concentrations and the TCDD concentration were found only for the samples from multipara mothers from Da Nang and Bien Hoa. The 1,2,3,4,6,7,8-HeptaCDF concentration was associated with the TCDD concentration for the samples from multipara mothers from Quan Tri. Significant associations were found between the concentrations of these four PCDD/F congeners and the TCDD concentrations in the samples from both primipara and multipara mothers in unsprayed areas, so decreased PCDD/F concentrations (particularly of PCDF congeners, which have shorter biological half-lives than PCDD congeners) in the tissues of mothers through breast feeding may increase associations between the concentrations of these four PCDD/F congeners and the TCDD concentrations. Advanced statistical analysis will be necessary in future to assess effects of breast feeding on the associations among PCDD/F concentrations.

4.2. Dioxin Concentrations in Breast Milk from Mothers Exposed to Dioxins in Countries Other Than Vietnam

4.2.1. Dioxin Levels in Breast Milk from Mothers from European Countries

An explosion of a herbicide manufacturing plant in 1976 caused TCDD contamination in Seveso, Italy. Many residents of the surrounding area were exposed to TCDD, and TCDD concentrations of <10–56,000 pg/g were found in serum from people living near the plant [24]. Weiss et al. determined PCDD/F concentrations in breast milk from nursing mothers in Seveso and reference areas in central Milan (urban control) and a rural area in Lombardy in 2000–2001, 25 years after the explosion [25]. The median concentrations of 17 PCDD/F congeners and the TEQ-PCDD/F concentrations that were found in the samples from Seveso, Milan, and Lombardy are shown in Table 6. The TCDD concentrations were more than two-times higher in the samples from Seveso than the samples from Milan and Lombardy, but the 1,2,3,6,7,8-HexaCDD and 1,2,3,4,6,7,8-HeptaCDD concentrations were lower in the samples from Seveso than the samples from Milan and Lombardy. These results indicated that TCDD concentrations were particularly high in the tissues of residents of Seveso even 25 years after the accident.

The Duisburg birth cohort study involved 169 mother–infant pairs recruited at childbirth in the highly industrialized Duisburg area in Germany in 2000–2003 [26]. The PCDD/F concentrations in breast milk samples were determined, and the median concentrations of 17 PCDD/F congeners are shown in Table 6. The congener profiles were similar to the congener profiles for the samples from Milan analyzed in a previous study [24] but the concentrations of the PCDD congeners except TCDD were slightly higher in the Duis-burg samples than the Milan samples. This suggested that industrialization may increase the concentrations of all PCDD/F congeners, particularly 1,2,3,6,7,8-HexaCDD, 1,2,3,4,6,7,8-HeptaCDD, OctaCDD, and 2,3,4,7,8-pentachlorodibenzofuran (PentaCDF), in human tissues. Compared with these congener concentrations in Da Nang and Quang Tri, three PCDD congener concentrations were similar in Duisburg area in Germany. However, the concentrations of 1,2,3,4,7,8-HexaCDF and 1,2,3,4,6,7,8-HeptaCDF were lower in Duisburg. These results suggest high 1,2,3,4,7,8-HexaCDF and 1,2,3,4,6,7,8-HeptaCDF concentrations may be characteristic in Da Nang and Quang Tri in middle Vietnam.

Table 6. Median levels of PCDD/F congeners and TEQ-PCDD/Fs in breast milk from various countries.

	Italy [21]			Germany [22]	Spain [23]	China [24]	Japan [25]	Vietnam [14]
	Seveso	**Milan**	**Lombardy**	**Duisburg**	**Catalonia**	**Tianjin**	**Hokuriku**	**PhuCat**
	n = 12	*n* = 12	*n* = 12	*n* = 169	*n* = 20	*n* = 20	*n* = 75	*n* = 16
PCDD congeners (pg/g lipid)								
2,3,7,8-TetraCDD	3.7	1.6	1.4	1.5	0.3	0.8	0.8	2.1
1,2,3,7,8-PentaCDD	2.4	3.6	3.3	4.0	0.8	1.6	4.2	5.7
1,2,3,4,7,8-HexaCDD	1.3	1.8	1.7	2.7	0.5	0.8	1.4	2.5
1,2,3,6,7,8-HexaCDD	7.3	9.5	8.6	11.6	2.6	2.1	13.7	10.0
1,2,3,7,8,9-HexaCDD	1.0	1.4	1.3	2.3	0.5	0.6	1.9	3.4
1,2,3,4,6,7,8-HeptaCDD	7.3	11.5	10.2	12.6	2.8	3.0	6.8	17.1
OctaCDD	38.6	49.5	50.2	70.4	15.7	20.5	56.2	80.9
PCDF congeners (pg/g lipid)								
2,3,7,8-TetraCDF	nd	nd	nd	0.3	0.2	0.7	1.0	0.7
1,2,3,7,8-PentaCDF	nd	0.2	nd	0.2	0.1	0.6	0.3	2.4
2,3,4,7,8-PentaCDF	5.9	8.9	7.3	10.1	1.9	5.6	8.1	7.6
1,2,3,4,7,8-HexaCDF	2.7	3.3	2.8	2.7	0.6	2.9	2.3	18.7
1,2,3,6,7,8-HexaCDF	1.7	2.4	1.9	2.4	0.6	2.2	2.3	10.6
1,2,3,7,8,9-HexaCDF	nd	nd	nd	1.0	0.4	nd	1.5	0.5
2,3,4,6,7,8-HexaCDF	0.6	0.9	0.7	0.1	nd	1.1	nd	1.7
1,2,3,4,6,7,8-HeptaCDF	0.7	1.4	0.7	1.9	0.9	1.0	1.2	19.8
1,2,3,4,7,8,9-HeptaCDF	nd	nd	nd	0.1	nd	nd	nd	2.1
OctaCDF	nd	nd	nd	0.3	nd	nd	0.6	1.3
TEQs (pg/g lipid)								
TEQ-PCDD/Fs	10.7	11.7	10.3	13.3	2.3	5.0	11.8	16.1

n: number of subjects; nd: not detected.

A hazardous waste incinerator was built in Tarragona County, Catalonia, Spain, in 1999. PCDD/F concentrations in breast milk from primipara mothers living near the incinerator were determined several times from 2002 onward [27]. The PCDD/F congener profiles in breast milk collected in 2017 (the latest report) are shown in Table 6. The concentrations of all PCDD/F congeners were low, and the median TEQ-PCDD/F was only 2.3 (pg-TEQ/g lipid) [27]. The authors stated that the decreasing PCDD/F levels were associated with decreasing dietary intakes of fatty foods such as fish and meat and that changes in environmental contamination had not affected human exposure to PCDD/Fs owing to the effective management of waste treatment and incineration processes [27].

4.2.2. Dioxin Levels in Breast Milk from Asian Countries

Sun et al. collected milk samples from mothers in three areas in China, Shijiazhuang (an industrial city), Tianjin (a large port city), and Yantai (a coastal city with a strong fishing industry), in 2006–2007 and determined the PCDD/F concentrations in the samples [28]. The median PCDD/F congener and TEQ-PCDD/F levels found in the Tianjin samples are shown in Table 6 because the concentrations were higher than the concentrations in the samples from the other areas. The PCDD/F congener and TEQ-PCDD/F concentrations in the Tianjin samples were generally lower than the concentrations found in samples from Lombardy in Italy in the study mentioned above [24] and similar to the concentrations found in samples from Catalonia in Spain in another study mentioned above [27]. The median levels of PCDD/F congener and TEQ-PCDD/F in unsprayed areas in Vietnam are also similar to levels in the Tianjin samples, suggesting PCDD/Fs contamination in Tianjin in China, Catalonia in Spain, and unsprayed areas in Vietnam may be at background levels.

The concentrations of most PCDD/F congener in breast milk from Hokuriku on the Japanese coast [29] were much higher than the concentrations found in the Tianjin samples [28] and similar to the concentrations found in the Milan samples [24], although the 1,2,3,6,7,8-HexaCDD concentrations were higher and the 1,2,3,4,6,7,8-HeptaCDD concentrations were lower in the Japanese samples than the Tianjin and Milan samples [24].

In 2008, Manh et al. determined the PCDD/F concentrations in breast milk from communities near Phu Cat airbase in Vietnam [18]. Phu Cat airbase, an important US military airbase during the Vietnam War, was used to store 17000 containers of Agent Orange, 9000 containers of Agent White, and 2900 containers of Agent Blue for the Ranch Hand mission [23]. The median PCDD/F congener concentrations in the samples are shown in Table 6. The PCDD/F concentrations were similar to the concentrations we found in the Da Nang samples (high TCDD, 1,2,3,6,7,8-HexaCDD, 1,2,3,4,6,7,8-HeptaCDD, 1,2,3,4,7,8-HexaCDF, and 1,2,3,4,6,7,8-HeptaCDF concentrations). The 1,2,3,7,8-pentachlorodibenzo-p-dioxin (PentaCDD) concentrations were higher than the concentrations found in samples from locations near other airbases in Vietnam, and this caused the TEQ-PCDD/F levels to be high. The PCDD/F congener patterns, particularly the high TCDD, 1,2,3,4,7,8-HexaCDF, and 1,2,3,4,6,7,8-HeptaCDF concentrations, were different from the patterns for samples from other countries, suggesting that PCDD/F exposure caused by herbicide spraying increased not only the TCDD concentrations but also the concentrations of other congeners such as 1,2,3,4,7,8-HexaCDF and 1,2,3,4,6,7,8-HeptaCDF in human tissues.

In Bien Hoa, Vietnam, Schecter et al. [13] reported strikingly high TCDD concentrations ranging from 333–1832 pg/g lipid in breast milk samples collected in 1970 (the last period of the Vietnam War). They also collected milk samples in subsequent years and reported decreasing TCDD concentrations year by year, as 133–266 pg/g lipid in 1973 and 2.1–11 pg/g lipid in 1985–1988 [13]. They did not show concentrations of PCDD/F congeners other than TCDD in breast milk samples, but the concentrations in their breast milk samples can be estimated to be not so high from their concentrations in blood samples collected in the same area [13]. Dwernychuk et al. also determined PCDD/F concentrations in breast milk collected in A So village in Aluoi Valley on central Vietnam in 1999 and reported 1.4–19 pg/g lipid for TCDD [14]. They also showed high total HeptaCDD (3.0–230 pg/g lipid) and total OCDD (28–1800 pg/g lipid), but concentrations of PCDF congeners were generally lower [14]. Because of differences in methodology of PCDD/Fs analysis, comparisons of concentrations between the samples collected before 2000 and in our samples collected after 2000 are not easy to conduct. However, increased concentrations of PCDF congeners observed in our samples might be specific to samples collected 40 years after the cessation of herbicide spraying.

4.3. Dioxins in Breast Milk and Other Markers of Exposure to PCDD/Fs

The concentrations of PCDD/Fs in breast milk are often used to indicate exposure of humans to dioxins because the high fat content of breast milk allows reliable measurements to be made. However, PCDD/F concentrations in maternal blood are better for estimating the exposure of a fetus to dioxins during pregnancy, which is associated with adverse health effects of the fetus and infants. The relationships between PCDD/F concentrations in breast milk and maternal blood have therefore been studied.

As part of the Duisburg birth cohort study, Wittsiepe et al. determined the concentrations of PCDD/Fs in blood and breast milk collected simultaneously from some mothers [26]. The median high-chlorinated (hexa-, penta-, and octa-chlorinated) PCDD/F congener concentrations were higher in the blood samples than the breast milk samples, although the low-chlorinated (tetra- and penta-chlorinated) PCDD/F congener concentrations were similar. The concentrations of the high-chlorinated PCDD/F congeners (except for the 1,2,3,4,7,8,9-HeptaCDF concentrations) in the blood and milk samples correlated well ($\gamma = 0.62$–0.94) [26].

Correlations between PCDD/F congener concentrations in cord blood and breast milk from mothers in the Bien Hoa birth cohort study in 2012 were determined, and good correlations were found for the low-chlorinated PCDD/F congeners, particularly TCDD, in the cord blood and breast milk samples [30].

The results of the studies described above indicate that PCDD/F concentrations in breast milk reflect maternal exposure to dioxins during pregnancy and are good markers for epidemiological studies of the effects of dioxins on infant health, particularly in Vietnam. TCDD is an important PCDD/F congener and PCDD/F congener analysis is necessary. However, drawing enough blood for dioxin congener analysis from pregnant women is very difficult in Vietnam.

Tolerable daily intake (TDI) of dioxins (1–4 pg-TEQ/kg body weight/day) is determined by the WHO [31] for prevention of adverse health effects with lifetime exposure. During early infancy, breast milk is a main source of postnatal dioxin exposure and estimated daily intake (EDI) of dioxins through breast milk for infants is calculated and compared with TDI values. In the present survey, mean EDI values or Quang Tri, Da Nang, and Bien Hoa were 57.3, 90.3, and 68.4 pg-TEQ/kg body weight/day and significantly higher than that for unsprayed areas (23.5 pg-TEQ/kg body weight/day). Even for unsprayed areas, the EDI value was above the TDI value, which was a similar level compared with countries exposed to dioxins at background levels of PCDD/Fs such as New Zealand [32] and Hungary [33]. When the 95th percentile value of EDI for unsprayed areas is set as cut-off value of high EDI, 60 percent of infants in Quang Tri, 90 percent of infants in Da Nang, and 73 percent of infants in Bien Hoa showed high EDI in the present analysis, suggesting large numbers of infants are at risk for potential adverse health effects due to PCDD/Fs exposure in sprayed areas in Vietnam.

However, no significant difference of mean EDI values between Quang Tri and Bien Hoa was shown, even if there was significant difference of TCDD concentrations originating from herbicides used during the Vietnam War. Therefore, determination of PCDD/F congener concentrations in breast milk samples may be necessary to evaluate PCDD/Fs exposure in the sprayed areas in Vietnam.

5. Conclusions

(1) The concentrations of almost all of the PCDD/F congeners were significantly higher in the samples from the herbicide-sprayed areas (Quang Tri, Da Nang, and Bien Hoa) than the unsprayed areas. (2) Different risks of high PCDD/F concentrations were found for the three herbicide-exposed areas. The risk of high TCDD concentrations was high for Bien Hoa, the risks of high TCDD and 1,2,3,6,7,8-HexaCDD concentrations were high for Da Nang, and the risk of high 1,2,3,4,6,7,8-HeptaCDD concentrations was high for Quan Tri. High 1,2,3,4,7,8-HexaCDF and 1,2,3,4,6,7,8-HeptaCDF concentrations were found only in samples from Da Nang and Quang Tri. (3) The associations between the TCDD and 1,2,3,4,7,8-HexaCDF and 1,2,3,4,6,7,8-HeptaCDF concentrations were different for samples from primipara and multipara mothers, suggesting that breast feeding affected PCDF concentrations more than PCDD concentrations. Advanced statistical analysis will be required in future to assess the characteristic PCDD/F congener profiles in breast milk samples from herbicide sprayed areas of Vietnam.

Author Contributions: Conceptualization, M.N. and H.N. (Hisao Nishijo); methodology, T.P.-T., H.N. (Hideaki Nakagawa) and M.N.; investigation, T.P-T., T.N.P., N.N.T. and H.T.V.; data curation, M.N.; writing—original draft preparation, M.N.; writing—review and editing, M.N. and H.T.V.; funding acquisition, M.N. and H.N. (Hideaki Nakagawa) All authors have read and agreed to the published version of the manuscript.

Funding: This work was partly supported by the Ministry of Education, Science, Sports and Culture with a Grant-in-Aid for Scientific Research (17H04665). The funders had no role in the design of the study; in the collection, analyses, or interpretation of data; in the writing of the manuscript, or in the decision to publish the results.

Institutional Review Board Statement: The study was conducted according to the guidelines of the Declaration of Helsinki and approved by the Institutional Review Boards of the Health Department of Da Nang City and the Kanazawa Medical University (No. E-187, September 2013).

Informed Consent Statement: Written informed consent was obtained from all subjects involved in the study.

Data Availability Statement: The data presented in this study are available on request to the corresponding author.

Conflicts of Interest: The authors declare no conflict of interest.

References

1. National Academy of Science. *The Effects of Herbicides in South Vietnam*; National Academy of Science Press: Washington, DC, USA, 1974; pp. S1–S6.
2. Stellman, J.M.; Stellman, S.D.; Christian, R.; Weber, T.; Tomasallo, C. The extent and patterns of usage of Agent Orange and other herbicides in Vietnam. *Nature* **2003**, *422*, 681–687. [CrossRef] [PubMed]
3. Schecter, A. A selective historical review of congener specific human tissue measurements as sensitive and specific biomarkers of exposure to dioxins and related compounds. *Environ. Health Perspect.* **1998**, *106*, 737–742. [PubMed]
4. Tai, P.T.; Nishijo, M.; Kido, T.; Nakagawa, H.; Maruzeni, S.; Naganuma, R.; Anh, N.T.; Morikawa, Y.; Luong, H.V.; Anh, T.H.; et al. Dioxin concentrations in breast milk of Vietnamese nursing mothers: A survey four decades after the herbicide spraying. *Environ. Sci. Technol.* **2011**, *45*, 6625–6632. [CrossRef]
5. Nghi, T.N.; Nishijo, M.; Manh, H.D.; Tai, P.T.; Van Luong, H.; Anh, T.H.; Thao, P.N.; Trung, N.V.; Waseda, T.; Nakagawa, H.; et al. Dioxins and nonortho PCBs in breast milk of vietnamese mothers living in the largest hot spot of dioxin contamination. *Environ. Sci. Technol.* **2015**, *49*, 5732–5742. [CrossRef]
6. Dwernychuk, L.W. Dioxin hot spots in Vietnam. *Chemosphere* **2005**, *60*, 998–999. [CrossRef] [PubMed]
7. The Office of the Vietnam National Steering Committee 33; Hatfield Consultants. *Comprehensive Assessment of Dioxin Contamination in Da Nang Airport, Vietnam: Environmental Levels, Human Exposure and Options for Mitigating Impacts*; Final Report; Hatfield Consultants: North Vancouver, BC, Canada, 2009; V7P 0A3.
8. The Office of the Vietnam National Steering Committee 33; Hatfield Consultants. *Environmental and Human Health Assessment of Dioxin Contamination at Bien Hoa Airbase, Viet Nam. Final Report*; Hatfield Consultants: North Vancouver, BC, Canada, 2011; V7P 0A3.
9. Pirkle, J.L.; Pirkle, J.L.; Wolfe, W.H.; Patterson, D.G.; Needham, L.L.; Michalek, J.E.; Miner, J.C.; Peterson, M.R.; Phillips, D.L. Estimates of the half-life of 2,3,7,8-tetrachlorodibenzo-p-dioxin in Vietnam Veterans of Operation Ranch Hand. *J. Tox. Environ. Health* **1989**, *27*, 165–171. [CrossRef]
10. Birnbaum, L.S. The mechanism of dioxin toxicity: Relationship to risk assessment. *Environ. Health Perspect* **1994**, *102* (Suppl. 9), 157–167. [CrossRef]
11. Ogawa, I. Half-life of each dioxin and PCB congener in the human body. *Organohalogen Compd.* **2004**, *66*, 3329–3337.
12. Tawara, K.; Nishijo, M.; Maruzeni, S.; Nakagawa, H.; Kido, T.; Naganuma, R.; Suzuki, H.; Nhu, D.D.; Hung, N.N.; Thom, L.T.H. Residual congener pattern of dioxins in human breast milk in southern Vietnam. *Chemosphere* **2011**, *84*, 979–986. [CrossRef]
13. Schecter, A.; Dai, L.C.; Päpke, O.; Prange, J.; Constable, J.D.; Mastuda, M.; Thao, V.D.; Piscac, A. Recent dioxin contamination from Agent orange in residents of a southern Vietnam city. *J. Occup Environ. Med.* **2001**, *43*, 435–443. [CrossRef]
14. Dwernychuk, L.W.; Cau, H.D.; Hatfield, C.T.; Boivin, T.G.; Hung, T.M.; Dung, P.T.; Thai, N.D. Dioxin reservoirs in southern Viet Nam—A legacy of agent orange. *Chemosphere* **2002**, *47*, 117–137. [CrossRef]
15. Tai, P.T.; Nishijo, M.; Anh, N.T.; Maruzeni, S.; Nakagawa, H.; Van Luong, H.; Anh, T.H.; Honda, R.; Kido, T.; Nishijo, H. Dioxin exposure in breast milk and infant neurodevelopment in Vietnam. *Occup. Environ. Med.* **2013**, *70*, 656–662. [CrossRef] [PubMed]
16. Pham, N.T.; Nishijo, M.; Pham, T.T.; Tran, N.N.; Le, V.Q.; Tran, H.A.; Phan, H.A.V.; Nishino, Y.; Nishijo, H. Perinatal dioxin exposure and neurodevelopment of 2-year-old Vietnamese children in the most contaminated area from Agent Orange in Vietnam. *Sci. Total Environ.* **2019**, *678*, 217–226. [CrossRef]
17. Saito, K.; Nhu, D.D.; Suzuki, H.; Kido, T.; Naganuma, R.; Sakakibara, C.; Tawara, K.; Nishijo, M.; Nakagawa, H.; Kusama, K.; et al. Association between dioxin concentrations in breast milk and food group intake in Vietnam. *Environ. Health Prev. Med.* **2010**, *15*, 48–56. [CrossRef] [PubMed]
18. Manh, H.D.; Kido, T.; Okamoto, R.; Xianliang, S.; Viet, N.H.; Nakano, M.; Tai, P.T.; Maruzeni, S.; Nishijo, M.; Nakagawa, H.; et al. The relationship between dioxins and salivary steroid hormones in Vietnamese primiparae. *Environ. Health Prev. Med.* **2013**, *18*, 221–229. [CrossRef]
19. Kido, T.; Honma, S.; Nhu, D.D.; Manh, H.D.; Van Tung, D.; Liang, S.X.; Anh, L.T.; Okamoto, R.; Maruzeni, S.; Nakagawa, H.; et al. Inverse association of highly chlorinated dioxin congeners in maternal breast milk with dehydroepiandrosterone levels in three-year-old Vietnamese children. *Sci. Total Environ.* **2016**, *550*, 248–255. [CrossRef] [PubMed]
20. Tawara, K.; Honda, R.; Nishijo, M.; Nakagawa, H. Pretreatment procedure of dioxin analysis for a small volume of human breast milk. *J. Kanazawa Med. Univ.* **2003**, *28*, 17–25.

21. Pham, N.T.; Nishijo, M.; Nghiem, T.T.G.; Pham, T.T.; Tran, N.N.; Le, V.Q.; Vu, T.H.; Tran, H.A.; Phan, H.A.V.; Do, Q.; et al. Effects of perinatal dioxin exposure on neonatal electroencephalography (EEG) activity of the quiet sleep stage in the most contaminated area from Agent Orange in Vietnam. *Int. J. Hyg. Environ. Health* **2021**, *232*, 113661. [CrossRef]
22. Van den Berg, M.; Birnbaum, L.S.; Denison, M.; De Vito, M.; Farland, W.; Feeley, M.; Fiedler, H.; Hakansson, H.; Hanberg, A.; Haws, L.; et al. The 2005 World Health Organization reevaluation of human and mammalian toxic equivalency factors for dioxins and dioxin-like compounds. *Toxicol. Sci.* **2006**, *93*, 223–241. [CrossRef]
23. Young, A.L. *The History, Use, Disposition and Environmental Fate of Agent Orange*; Springer: New York, NY, USA, 2009.
24. Mocarelli, P.; Needham, L.L.; Marocchi, A.; Patterson, D.G., Jr.; Brambilla, P.; Gerthoux, P.M.; Meazza, L.; Carreri, V. Serum concentrations of 2,3,7,8-tetrachlorodibenzo-p-dioxin and test results from selected residents of Seveso, Italy. *J. Toxicol. Environ. Health* **1991**, *32*, 357–366. [CrossRef]
25. Weiss, J.; Papke, O.; Bignert, A.; Jensen, S.; Greyerz, E.; Agostoni, C.; Besana, R.; Riva, E.; Giovannini, M.; Zetterstrom, R. Concentrations of dioxins and other organochlorines (PCBs, DDTs, HCHs) in human milk from Seveso, Milan and a Lombardian rural area in Italy: A study performed 25 years after the heavy dioxin exposure in Seveso. *Acta Paediatr.* **2003**, *92*, 467–472. [CrossRef]
26. Wittsiepe, J.; Furst, P.; Schrey, P.; Lemm, F.; Kraft, M.; Eberwein, G.; Winneke, G.; Wilhelm, M. PCDD/F and dioxin-like PCB in human blood and milk from German mothers. *Chemosphere* **2007**, *67*, S286–S294. [CrossRef]
27. Schuhmacher, M.; Mari, M.; Nadal, M.; Domingo, J.L. Concentrations of dioxins and furans in breast milk of women living near a hazardous waste incinerator in Catalonia, Spain. *Environ. Int.* **2019**, *125*, 334–341. [CrossRef]
28. Sun, S.; Zhao, J.; Leng, J.; Wang, P.; Wang, Y.; Fukatsu, H.; Liu, D.; Liu, X.; Kayama, F. Levels of dioxins and polybrominated diphenyl ethers in human milk from three regions of northern China and potential dietary risk factors. *Chemosphere* **2010**, *80*, 1151–1159. [CrossRef]
29. Tawara, K.; Nishijo, M.; Honda, R.; Maruzeni, S.; Seto, T.; Kido, T.; Saito, S.; Nakagawa, H. Effects of maternal dioxin exposure on newborn size at birth among Japanese mother-infant pairs. *Environ. Health Prev. Med.* **2009**, *14*, 88–95. [CrossRef]
30. Boda, H.; Nghi, T.N.; Nishijo, M.; Thao, P.N.; Tai, P.T.; Van Luong, H.; Anh, T.H.; Morikawa, Y.; Nishino, Y.; Nishijo, H. Prenatal dioxin exposure estimated from dioxins in breast milk and sex hormone levels in umbilical cord blood in Vietnamese newborn infants. *Sci. Total Environ.* **2018**, *615*, 1312–1318. [CrossRef]
31. van Leeuwen, F.X.; Feeley, M.; Schrenk, D.; Larsen, J.C.; Farland, W.; Younes, M. Dioxins: WHO's tolerable daily intake (TDI) revisited. *Chemosphere* **2000**, *40*, 1095–1101. [CrossRef]
32. 't Mannetje, A.; Coakley, J.; Bridgen, P.; Smith, A.H.; Read, D.; Pearce, N.; Douwes, J. Estimated infant intake of persistent organic pollutants through breast milk in New Zealand. *N. Z. Med. J.* **2014**, *127*, 56–68.
33. Vigh, É.; Colombo, A.; Benfenati, E.; Håkansson, H.; Berglund, M.; Bódis, J.; Garai, J. Individual breast milk consumption and exposure to PCBs and PCDD/Fs in Hungarian infants: A time-course analysis of the first three months of lactation. *Sci. Total Environ.* **2013**, *449*, 336–344. [CrossRef]

toxics

MDPI

Article

Histopathological Alterations in the Livers of Chronic Hepatitis Patients Exposed to Agent Orange/Dioxin in Vietnam

Phu Quang Pham [1], Vuong Ba Nguyen [2,*], Tai The Pham [3], Nhuong Xuan Duong [1], Ha Trong Nguyen [2], Quang Van Ha [2], Thuan Duc Nguyen [2], Tuan Minh Hoang [1], Dong Tien Dinh [1], Quynh Thi Nhu Tran [4], Linh Kim Bui [5], Thuy Thi Vu [6], Manh Van Phan [7], Tuan Minh Luong [8], Khanh Nguyen [9], Dung Anh Vu [10] and Thao Ngoc Pham [11]

1 Department of Gastroenterology, 103 Military Hospital, Vietnam Military Medical University, Hanoi 12108, Vietnam; bsphu79@yahoo.com (P.Q.P.); nhuongdx171@yahoo.com.vn (N.X.D.); drhoangtuanhvqy@gmail.com (T.M.H.); bsdonga1@gmail.com (D.T.D.)
2 Department of Haematology, Toxicology, Radiation and Occupation, 103 Military Hospital, Vietnam Military Medical University, Hanoi 12108, Vietnam; bshaa7103@gmail.com (H.T.N.); haquangss@gmail.com (Q.V.H.); nducthuanbv103@gmail.com (T.D.N.)
3 Biomedical and Pharmaceutical Research Centre, Vietnamese Military Medical University, Hanoi 12108, Vietnam; taithuy@kanazawa-med.ac.jp
4 General Internal Medicine Department, National Hospital of Endocrinology, Hanoi 11500, Vietnam; tranquynh1785@gmail.com
5 Department of Military Epidemiology, Vietnam Military Medical University, Hanoi 12108, Vietnam; builinhhvqy@gmail.com
6 Department of Gastroenterology, Huu Nghi Hospital, Hanoi 11600, Vietnam; thuyvubvhuunghi@gmail.com
7 Department of Military Hygiene, Vietnam Military Medical University, Hanoi 12108, Vietnam; bsphanvanmanh@gmail.com
8 Army Medical Department, Ministry of Defence, Hanoi 11100, Vietnam; luongtuan27@gmail.com
9 Department of Thoracic Surgeon, 103 Military Hospital, Vietnam Military Medical University, Hanoi 12108, Vietnam; drkhanh90@gmail.com
10 Department of Orthopedic Surgeon, 103 Military Hospital, Vietnam Military Medical University, Hanoi 12108, Vietnam; surgeonvuanhdung@gmail.com
11 Department of Functional Diagnosis, 103 Military Hospital, Vietnam Military Medical University, Hanoi 12108, Vietnam; phamngocthaovmmu@gmail.com
* Correspondence: bavuongsang@gmail.com; Tel.: +84-912340674

Citation: Pham, P.Q.; Nguyen, V.B.; Pham, T.T.; Duong, N.X.; Nguyen, H.T.; Ha, Q.V.; Nguyen, T.D.; Hoang, T.M.; Dinh, D.T.; Tran, Q.T.N.; et al. Histopathological Alterations in the Livers of Chronic Hepatitis Patients Exposed to Agent Orange/Dioxin in Vietnam. *Toxics* **2022**, *10*, 315. https://doi.org/10.3390/toxics10060315

Academic Editors: Anita K. Patlolla and Christopher J. Martyniuk

Received: 19 April 2022
Accepted: 6 June 2022
Published: 10 June 2022

Abstract: We investigated changes in some laboratory indices and the liver histology of chronic hepatitis patients who were exposed to dioxin. In 2014, we collected liver biopsy samples for histopathological examination from 33 chronic hepatitis patients living around the Da Nang Airbase, which is a dioxin-contaminated area due to the herbicide spraying in Vietnam. Dioxin exposure was measured by its levels in the blood. METAVIR classification was used to clarify the liver fibrosis stage. Laboratory tests included ten biochemical and six hematological indices that were measured in the blood. A regression linear model and binary logistic regression were used for data analysis. The observed alterations in the liver at the histological level mainly comprised hydropic degenerative hepatocytes, lymphocytes and polynuclear leukocytes surrounding the liver cells and granular and lipoic degeneration. In addition, increased TCDD levels were associated with increasing aminotransferase (AST), alanine aminotransferase, protein and total bilirubin levels and liver fibrosis stage. Similarly, increased TEQ-PCDD/Fs levels were associated with higher levels of AST and protein and liver fibrosis stage. In conclusion, dioxin exposure altered the liver histology and increased some biochemical marker indices and the liver fibrosis stage of chronic hepatitis patients living in dioxin-contaminated areas in Da Nang, Vietnam.

Keywords: dioxin; histopathological change; liver enzyme; liver damage; Vietnam

1. Introduction

During the Vietnam War from 1961 to 1972, the US Army carried out an herbicide spraying campaign with a huge amount of Agent Orange, which contained 2,3,7,8-tetrachlorodibenzo-p-dioxin (TCDD), in the southern areas of Vietnam. The impacts of the dioxin that originated from Agent Orange have been reported in both environmental and human health, even though the campaign ended over 50 years ago. Dioxin is absorbed into the body in various ways, including via the mucosal, skin, respiratory and gastrointestinal tracts; then, the circulatory system helps to distribute the dioxin around the other organs in the body [1]. Dioxin is insoluble in water and once it is absorbed into the bloodstream, it only exists in the blood for a short time before accumulating in fatty tissues and the liver. Moreover, the liver plays an essential role in metabolism and in immobilizing and inactivating the internal and external toxins within the human body. Therefore, it has been suggested that the liver is more vulnerable to damage that is caused by toxification from environmental toxins, including Agent Orange.

In previous studies, it has been reported that the liver is the main target of the toxic effects that are caused by TCDD exposure [2–6]. Serdar et al., (2014) investigated the effects of polychlorinated biphenyls (PCBs) and organochlorine pesticides (OCGs), which have similar toxicity to dioxin, on blood biochemistry markers. They showed that increased exposure to PCBs and OCGs is associated with increased levels of aspartate aminotransferase (AST), alanine aminotransferase (ALT) and gamma-glutamyl transpeptidase (GGT) [7]. Increased liver enzyme levels, including transaminases and GGT, have also been reported among residents who were exposed to TCDD by the Seveso accident [2]. An increased prevalence of liver disease, including liver cirrhosis, has been associated with exposure to dioxin originating from Agent Orange in Vietnam [3–6]. In addition, microscopic examinations (histological examinations) have also reported that dioxin causes morphological lesions in hepatic parenchymal cells, such as parenchymal degeneration [8], and electron microscopical evaluations in experimental studies have reported nucleus, mitochondria, rough endoplasmic reticulum, cytostomes and bile canaliculus in the cytoplasm [9]. However, limited studies have investigated the effects of dioxin exposure on the human liver using microscopes or electron microscopes, except a study of patients exposed to PCBs and dioxin by the soot from the Binghamton State Office Building incident. They found that TCDD and PCBs induce morphologic alterations in liver cells, as indicated by changes in the hepatic parenchymal cell cytoplasm, endoplasmic reticulum or mitochondria of patients who showed mild liver enzyme levels, including GGT, AST and ALT. These alterations were similar to those seen in animals that were exposed to PCBs and dioxin via feeding, although they did not produce the same microscopical findings and the number of subjects in the study was small, with only three patients [10]. From these results, we hypothesized that alterations in liver morphology at the histological level among chronic hepatitis patients were caused by dioxin exposure from Agent Orange.

Therefore, in the present study, we investigated the associations between dioxin exposure and some biomarker indices, including liver enzyme levels, lipid metabolism, hematology and histological alterations in the livers of patients who were diagnosed with chronic hepatitis and exposed to dioxin that originated from Agent Orange in Vietnam.

2. Materials and Methods

2.1. Study Areas and Subjects

The Da Nang Airbase is located inside the city of Da Nang in a condemned neighborhood that was used to store herbicide in Vietnam and is considered to be a hot spot for dioxin contamination [11]. Hatfield Consultants monitor environmental hazards, including heavy metals and organochlorines, and reported that the TCDD concentrations in soil samples collected inside the airbase from December 2006 to January 2009 ranged from 858 to 361,000 pg/g dry weight. In addition, elevated dioxin concentrations have been found not only in the blood of individuals working on the airbase but also in that of residents living around the airbase [12,13].

A total of 40 patients living in the hot spot of dioxin contamination within Da Nang city were selected from 17 military hospitals from August 2014 to January 2015. The criteria for the subject recruitment were as follows: (i) patients volunteered to participate the survey; (ii) patients were in the age range of 18–70 years; (iii) the liver enzyme levels of the patients were continuously high or were high for intermittent periods of at least six months (indicated by live biopsies to diagnose chronic hepatitis) [14]; (iv) patients lived in the hot spot of dioxin contamination (Da Nang Airbase) for at least five years and the toxic equivalent (TEQ) of polychlorinated dibenzo-p-dioxins (PCDDs) and polychlorinated dibenzofurans (PCDFs) TEQ levels in their serum was higher than the background levels (>9.4 pg-TEQ/g lipid) [15]; (vi) patients were not diagnosed with chronic hepatitis due to other factors, including the hepatitis B and C viruses, alcohol, drugs and autoimmune diseases, and were contraindicated for liver biopsy [14]. The patients were diagnosed with chronic hepatitis based on histopathological findings: mono leukocytes (mainly lymphocytes) appearing in portal spaces and fibrosis imaging [16,17]. Of these, seven patients refused to participate in the liver biopsy procedure. Therefore, the final number of subjects in our data analysis was 33 patients.

Information about the patients was collected, including their age, gender, smoking status, alcohol consumption (alcohol drinking history) and the number of years that they had lived in dioxin contaminated areas until the recruitment time.

Written informed consent was obtained from all of the participants, according to a process that had been reviewed and approved by the Ethics Council of the Vietnam Military Medical University (Number: 33.13/11-15).

2.2. Dioxin Measurements

A 40 mL sample of whole blood was collected from each patient. Then, the samples were centrifuged to separate the serum and stored at −30 °C in the Vietnam Russia Tropical Center until analysis. Dioxin measurement was performed in the dioxin laboratory at the Chemical–Environment Sub-Institute of the Vietnam Russia Tropical Center. The laboratory has been recognized for compliance with ISO/IEC 17025: 2017 and has received VILAS 856. Firstly, the serum samples were added to the internal standards of the 13C12-PCDD/PCDF isotope and then, under a procedure of liquid–liquid extraction, to the hexane–ethanol mixture. The extraction was recovered by n-hexane solution. The PCDD/PCDF fractions were separated on a dedicated activated carbon column, which was followed by cleaning steps on a "multi-layer column" containing silica gel, acid-impregnated silica gel and alkaline silica gel. Finally, the PCDD/PCDF fractions were separated through an aluminum oxide column after adding a standard of the 13C12-PCDD isotope to determine recovery efficiency. Dioxin measurement was followed by high-resolution gas chromatography/high-resolution mass spectrometry (GC-MS) using gas chromatography (Agilent 7890A) and high-resolution mass spectrometry (AutoSpec Premier P834). Certified reference materials from the National Institute of Standards and Technology (NIST) were regularly analyzed for quality control. The detection limits for each congener (in picograms per gram of wet weight) in the serum were as follows: TCDD, 0.013; 1,2,3,7,8-pentachlorodibenzo-p-dioxin (1,2,3,7,8-PCDD), 0.013; 1,2,3,4,7,8-hexachlorodibenzo-p-dioxin (1,2,3,4,7,8-HCDD), 0.033; 1,2,3,6,7,8-hexachlorodibenzo-p-dioxin (1,2,3,6,7,8-HCDD), 0.033; 1,2,3,7,8,9-hexachlorodibenzo-p-dioxin (1,2,3,7,8,9-HCDD), 0.033; 1,2,3,4,6,7,8-heptachlorodibenzo-p-dioxin (1,2,3,4,6,7,8-HpCDD), 0.033; octachlorodibenzo-p-dioxin (OCDD), 0.033; 2,3,7,8-tetrachlorodibenzofuran (2,3,7,8-TCDF), 0.013; 1,2,3,7,8-pentachlorodibenzofuran (1,2,3,7,8-PCDF), 0.013; 2,3,4,7,8-pentachlorodibenzofuran (2,3,4,7,8-PCDF), 0.033; 1,2,3,4,7,8-hexachlorodibenzofuran (1,2,3,4,7,8-HCDF), 0.033; 1,2,3,6,7,8-hexachlorodibenzofuran (1,2,3,6,7,8-HCDF), 0.033; 1,2,3,7,8,9-hexachlorodibenzofuran (1,2,3,7,8,9-HCDF), 0.033; 2,3,4,6,7,8-hexachlorodibenzofuran (2,3,4,6,7,8-HCDF), 0.033; 1,2,3,4,6,7,8-heptachlorodibenzofuran (1,2,3,4,6,7,8-HCDF), 0.033; 1,2,3,4,7,8,9-heptachlorodibenzofuran (1,2,3,4,7,8,9-HpCDF), 0.033; octachlorodibenzofuran (OCDF), 0.033. The limit of quantitation for each dioxin congener (pg/g) in the serum was as follows: TCDD, 0.017; 1,2,3,7,8-PeCDD, 0.0833; 1,2,3,4,7,8-HxCDD, 0.0833;

1,2,3,6,7,8-HxCDD, 0.0833; 1,2,3,7,8,9-HxCDD, 0.0833; 1,2,3,4,6,7,8-HpCDD, 0.0833; OCDD, 0.1667; 2,3,7,8-TCDF, 0.017; 1,2,3,7,8-PeCDF, 0.0833; 2,3,4,7,8-PeCDF, 0.0833; 1,2,3,4,7,8-HxCDF, 0.0833; 1,2,3,6,7,8-HxCDF, 0.0833; 2,3,4,6,7,8-HxCDF, 0.0833; 1,2,3,7,8,9-HxCDF, 0.0833; 1,2,3,4,6,7,8-HpCDF, 0.0833; 1,2,3,4,7,8,9-HpCDF,: 0.0833; OCDF, 0.1667. The TEQs of the seven PCDD and ten PCDF congeners were calculated as the sum of all values, which were obtained by multiplying each congener concentration by its toxic equivalent factor from the WHO 2005-TEF [18].

2.3. Laboratory Tests

Ten biochemical and six hematological parameters were measured in the serum of each patient selected from 17 military hospitals in Da Nang, Vietnam. An amount of 2 mL of venous blood was withdrawn from each patient for biochemical testing. Liver enzymes, consisting of GGT, AST and ALT, and other biochemical markers, including protein, cholesterol, triglyceride, total bilirubin, glucose, urea and creatinine, in the serum were measured using the fully automated Hitachi 912 Chemistry Analyzer (Roche Diagnostics, Mannheim, Germany). All reagent kits were obtained from Roche Diagnostics, India. Similarly, 2 mL of venous blood was used to measure the six hematological indices, including red blood cell count, hemoglobin, leukocytes, neutrocytes, lymphocytes and platelets, using the DxH 600 Hematology Analyzer (Beckman coulter, Seattle, WC, USA).

2.4. Histopathological Examination

Liver histopathology plays a critical role in diagnosing and finding the causes of liver disease, as well as being the gold standard for diagnosing and assessing liver damage. We used 16-gauge biopsy needles (DeltaCut, Pajunk Medizintechnologie GmbH, Baden-Württemberg, Germany) with the needle length of the gun adjusted from 15 to 22 mm to collect the liver biopsy samples. The samples were selected for histopathology examination when they met the following criteria: the specimen size was >1.5 cm with six portal spaces and a diameter of 1.4 mm. Then, the specimens were fixed with 10% formol, cast with paraffin and cut into 4-µm slices to form the template before being stained with hematoxylin–eosin. The Olympus BX51 electron microscope (Olympus, Tokyo, Japan) was used to examine the biopsy samples.

The METAVIR scoring system is simple and the most commonly used system in clinical practice. According to METAVIR, there are five stages of fibrosis, as follows: F0, no fibrosis; F1, portal fibrosis without spaces; F2, portal fibrosis and several bridges; F3, fibrosis with multiple bridges or bridge fibrosis, F4, cirrhosis [19]. Figure 1 displays the characteristics of the F0, F1 and F2 stages of fibrosis.

2.5. Statistical Analysis

SPSS v. 21.0 for Windows (IBM Corp., Armonk, NY, USA) was used for the data analysis. The concentrations of TCDD and TEQ-PCDD/Fs and the laboratory indices were logarithmically transformed (base 10) to improve normality. The relationships between the TCDD and TEQ-PCDD/Fs levels and the laboratory indices were analyzed using a regression linear model after adjusting for covariates, including gender, age and smoking status. At this time, variables were selected as covariates if they correlated with at least one biochemical or hematological marker, based on a Pearson's correlation analysis, or if the groups differed significantly in at least one biochemical or hematological marker (Student's *t*-test, $p < 0.05$). Since the number of the patients who showed the F0 METAVIR grade was small in the present study, we divided the subjects into positive and negative METAVIR scores as follows: positive METAVIR score for those who showed liver fibrosis at stage F2 and negative METAVIR scores for those who showed liver fibrosis at stages F0 or F1. Then, the associations between dioxin exposure, as indicated by the TCDD and TEQ-PCDD/Fs levels, and positive METAVIR scores were analyzed using a binary logistic regression model after adjusting for the same confounding factors as above. A *p*-value of < 0.05 was considered statistically significant.

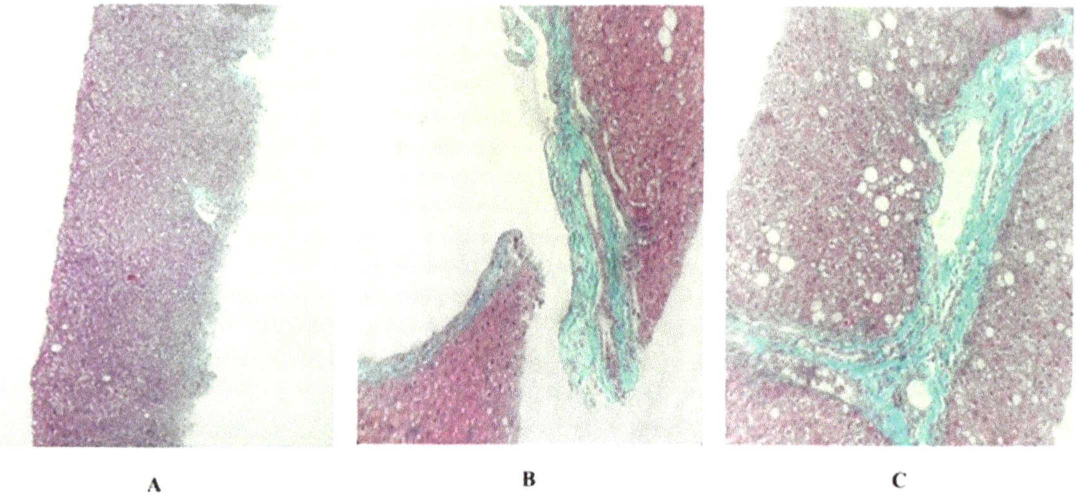

Figure 1. Histopathology of liver tissues under 10× microscope. The METAVIR fibrosis staging system (F0, F1, F2): (**A**) F0 (the core needle biopsy shows liver tissue with two central veins, including the infiltration of some chronic inflammatory cells, many hepatocytes with hydropic degeneration and a few hepatocytes with fatty degeneration, but there is no appearance of portal zones); (**B**) F1 (the core needle biopsy shows liver tissue with one portal zone and no central veins, including the infiltration of numerous chronic inflammatory cells, a strong development of connective tissue surround portal zone, the bile duct is expanded and many hepatocytes contain fatty degeneration, but there is no appearance of bridging); (**C**) F2 (the core needle biopsy shows liver tissue with one or two portal zones and probably one central vein, including the infiltration of numerous chronic inflammatory cells, a strong development of connective tissue surrounding portal zones, many bile ducts are expanded, many hepatocytes contain large fatty degeneration and there is probably the appearance of bridging between central vein and portal zone).

3. Results

3.1. Characteristics of the Subjects

The characteristics of the study subjects are shown in Table 1. The average age of the subjects was 46.3 years, with the rate of male subjects and patients who were smokers being 48.5% of total subjects. None of the subjects consumed alcoholic beverages. The mean length of time that patients lived in a hot spot for dioxin contamination was 26.0 years. The mean body mass index of the subjects was within the normal range (<25). For the METAVIR score, the percentage of patients with the F0 grade was the lowest, which accounted for 17.3% of the subjects. The percentage of patients with F1 and F2 grades were 42.4% and 30.3%, respectively. The geometrical mean of the TCDD and TEQ-PCDD/Fs levels in the blood was 15.7 pg/g lipid and 42.5 pg-TEQ/g lipid, which was found to be seven or eight time higher than those in unsprayed areas of Vietnam [20].

3.2. Associations between TCDD and TEQ-PCDD/Fs Exposure and Laboratory Indices

The associations between dioxin exposure, as indicated by the levels of TCDD and TEQ-PCDD/Fs in the blood, and the biochemical and hematological indices were analyzed using a regression linear model after adjusting for gender, age and smoking status. The results are illustrated in Table 2.

The TCDD exposure levels had significant and positive associations with AST, ALT, total bilirubin and protein levels ($p < 0.05$). Similarly, the TCDD exposure levels had borderline significant and positive associations with GGT levels. However, no significant associations between the TCDD exposure levels and the other biomarker indices were found (Table 2).

Table 1. Characteristics of the study subjects.

	Characteristic	Unit	Mean (SD), N [%]
Subjects			
	Age	year	46.3 (12.1)
	Ratio of Male	% male	16 [48.5%]
	Smoking Status	% smoker	16 [48.5%]
	Alcohol Consumption	% drinking alcohol	0 [0.0%]
	Length stayed in dioxin contaminated areas	Year	26.0 (10.6)
	BMI		21.7 (2.1)
METAVIR Score			
	F0	%	9 [17.3%]
	F1	%	14 [42.4%]
	F2	%	10 [30.3%]
Dioxin Concentration in Blood			
	TCDD	pg/g lipid	15.7 (3.7) *
	TEQ-PCDD/Fs	pg-TEQ/g lipid	42.5 (2.5) *

N, number of subjects; SD, standard deviation; BMI, body mass index; TCDD, 2,3,7,8-tetrachlorodibenzo-p-dioxin; PCDD/Fs, polychlorinated dibenzo-p-dioxins and polychlorinated dibenzo furans; TEQ, toxic equivalent quantity; *, geometrical mean and geometrical standard.

Table 2. Associations between TCDD and TEQ-PCDD/Fs concentration and biomarker indices.

Markers	N	TCDD β	TCDD 95% CI (Lower, Upper)	TCDD *p*	TEQ-PCDD/FS β	TEQ-PCDD/FS 95% CI (Lower, Upper)	TEQ-PCDD/FS *p*
Aspartate Aminotransferase (AST)	33	0.414	(0.071, 0.758)	0.020	0.393	(0.046, 0.741)	0.028
Alanine Aminotransferase (ALT)	33	0.359	(0.009, 0.709)	0.045	0.349	(−0.003, 0.702)	0.052
Gamma-Glutamyl Transpeptidase (GGT)	33	0.340	(−0.020, 0.699)	0.063	0.298	(−0.068, 0.664)	0.107
Total Bilirubin	33	0.389	(0.030, 0.748)	0.035	0.283	(−0.091, 0.656)	0.133
Glucose	33	−0.05	(−0.417, 0.318)	0.79	−0.14	(−0.502, 0.229)	0.45
Urea	33	0.031	(−0.318, 0.381)	0.855	−0.013	(−0.364, 0.337)	0.938
Creatinin	33	−0.011	(−0.284, 0.262)	0.933	0.062	(−0.211, 0.334)	0.648
Protein	33	0.421	(0.104, 0.738)	0.011	0.351	(0.021, 0.680)	0.038
Cholesterone	33	0.029	(−0.334, 0.392)	0.871	0.006	(−0.358, 0.370)	0.974
Triglyceride	33	0.043	(−0.334, 0.420)	0.816	0.104	(−0.272, 0.480)	0.576
Red Blood Cells	33	0.137	(−0.200, 0.474)	0.412	0.078	(−0.263, 0.418)	0.644
Hemoglobin	33	0.003	(−0.288, 0.294)	0.985	−0.024	(−0.316, 0.267)	0.866
Leukocytes	33	−0.031	(−0.420, 0.357)	0.870	−0.225	(−0.605, 0.155)	0.236
Neutrocytes	33	−0.201	(−0.582, 0.181)	0.291	−0.240	(−0.619, 0.139)	0.205
Lymphocytes	33	0.234	(−0.146, 0.614)	0.218	0.299	(−0.075, 0.673)	0.113
Platelets	33	−0.172	(−0.549, 0.206)	0.360	−0.127	(−0.508, 0.254)	0.500

TCDD, 2,3,7,8-tetrachlorodibenzo-p-dioxin; PCDD/Fs, polychlorinated dibenzo-p-dioxins and polychlorinated dibenzo furans; TEQ, toxic equivalent quantity; N, number of subjects; β, standardized beta; 95% CI, 95% confidence interval. Confounding factors: gender; age; smoking status.

Similarly, the TEQ-PCDD/Fs levels were significantly and positively associated with AST and protein levels ($p < 0.05$). Increased TEQ-PCDDs/Fs exposure levels had borderline significant associations with increased ALT levels ($p = 0.052$). No significant associations were found between the TEQ-PCDD/Fs exposure levels and GGT or the other biomarker indices measured in the present study (Table 2).

3.3. The Characteristics of Histopathological Damage

Chronic liver injury at the histological level was detected in almost all biopsy samples. The results are displayed in Table 3. The alterations of hydropic degenerative hepatocytes and lymphocytes and polynuclear leukocytes surrounding the liver cells were observed in all samples. Granular and lipoic degeneration was found in 32 of the total 33 samples, which accounted for 97% of the subjects. There were four cases (12.1%) that showed changes in hepatocytes eosinophils. Other indications of chronic liver disease, such as Mallory bodies, pigmentation, lipo-granuloma, megamitochondria and venous obstruction, were not observed in the present study (Table 3).

Table 3. Histopathological damage among chronic hepatitis patients who were exposed to dioxin.

Histopathological Damage	Yes N (%)	No N (%)
Granular Degeneration	32 (97.0)	1 (3.0)
Hydropic Degeneration	33 (100)	0 (0.0)
Lipoic Degeneration	32 (97.0)	1 (3.0)
Lipogranuloma	0 (0.0)	33 (100)
Lymphocytes and Polynuclear Leukocytes Surrounding the Liver Cells	33 (100)	0 (0.0)
Mallory Bodies	0 (0.0)	33 (100)
Pigmentation	0 (0.0)	33 (100)
Megamitochondria	0 (0.0)	33 (100)
Changes in Acidophil Hepatocyte	4 (12.1)	29 (87.9)
Venous Obstruction	0 (0.0)	33 (100)

N, number of subjects.

3.4. Associations between TCDD and TEQ-PCDD/Fs Exposure and METAVIR Scores

METAVIR scores were used to clarify the degree of liver fibrosis. The highest fibrosis level in the present study was F2, which accounted for 30.3% of the subjects. The percentage of F1 and F2 grades were 42.4% and 27.3%, respectively. No F3 and F4 grades were found in the samples from this group, suggesting that there were no samples that showed a fibrous expansion of portal areas with marked bridging, marked bridging with occasional nodules (incomplete cirrhosis) or cirrhosis [19] (Table 1). To find the effects of dioxin exposure on liver fibrosis level, the relationships between TCDD and TEQ-PCDD/Fs exposure levels and positive METAVIR scores were analyzed using a binary logistic regression model, which was adjusted for age, gender and smoking status. The results are shown in Table 4. The results showed that increased exposure levels of TCDD and TEQ-PCDD/Fs were significantly associated with increased odd ratios of positive fibrosis grades (TCDD, OR = 5.8 and $p = 0.007$; TEQ-PCDD/Fs, OR = 3.8 and $p = 0.021$) (Table 4).

Table 4. Relationships between TCDD and TEQ-PCDD/Fs exposure and positive METAVIR scores (F2).

	OR	SE	95% CI (Lower, Upper)	p
TCDD	5.9	0.7	(1.6, 21.4)	0.007
TEQ-PCDDs/Fs	3.9	0.6	(1.2, 12.1)	0.021

TCDD, 2,3,7,8-tetrachlorodibenzo-p-dioxin; TEQ, toxic equivalent quantity; PCDD/Fs, polychlorinated dibenzo-p-dioxins and polychlorinated dibenzo furans; OR, odd ratio; SE, standard error; 95% CI, 95% confidence interval. Covariates: age; gender; smoking status.

4. Discussion

In this study, we found that increased exposure to TCDD was significantly associated with increased serum AST and ALT levels. TCDD exposure levels also had borderline significant and positive associations with GGT levels. Similarly, increased exposure to TEQ-PCDD/Fs had significant or borderline significant associations with increased serum AST and ALT levels. These results suggest that dioxin exposure, as indicated by levels of

TCDD and TEQ-PCDD/Fs in the blood, increases serum liver enzyme levels, particularly AST and ALT.

In previous studies, investigations were conducted on the effects of PCBs and OCGs, which have similar toxicity to dioxin, on blood biochemistry markers during the National Health and Nutrition Examination Survey (NHANES 2003–2004) in the US. Serdar et al., (2014) reported that increased exposure to PCBs and OCGs is associated with increased AST, ALT and GGT levels [7]. Increased liver enzyme levels, including transaminases and GGT, were also reported among residents who were exposed to TCDD during the Seveso accident [2]. In a study in Germany, Triebig et al., (1998) recruited 76 former workers from a non-ferrous metal recycling facility who showed elevated levels of TEQ-PCDD/Fs (median = 42 ppt, range = 13–281). The authors reported significant and positive associations between dioxin exposure and liver enzyme levels, indicated by alanine aminotransferase, albeit no such associations were observed with serum cholesterol levels or high-density lipoprotein cholesterol (HDL) in this study [21]. In contrast, these associations were not found in other studies [22,23]. In a group of 138 former chemical workers who had been potentially exposed to TCDD following the 1953 trichlorophenol autoclave accident, there were no significant associations between TCDD exposure and liver enzyme (AST, ALT and GGT), serum glucose, lipid metabolism (cholesterol, triglyceride, HDL and low-density lipoprotein cholesterol (LDL)), hematology or coagulation parameters, except platelets [23]. In this study, we also found that there were no associations between TCDD and TEQ-PCDD/Fs exposure and hematological parameters (glucose, urea and creatinine) or lipid parameters (serum cholesterol and triglyceride levels); however, HDL and LDL levels were not measured in the present study. Positive associations were found between exposure to TCDD and TEQ-PCDD/Fs and plasma protein and total bilirubin levels. In addition, alterations in liver enzyme, lipid metabolism, total protein, bilirubin, blood cell, hemoglobin and plasma protein levels were induced by TCDD exposure, which was most probably associated with food consumption and dietary intake [24]. Therefore, statistical associations between TCDD and TEQ-PCDD/Fs exposure that originated from Agent Orange and liver enzyme parameters (glucose and blood cells), lipid parameters (cholesterol and triglyceride) and other biomarker indices in serum should be re-evaluated in a follow-up study after controlling food intake.

In addition, in the present study, we found alterations in liver morphology at the histological level, as indicated by granular degeneration, hydropic degeneration, lipoic degeneration and lymphocytes and polynuclear leukocytes surrounding the liver cells, in most biopsy samples from chronic hepatitis patients who were exposed to high levels of dioxin. Our report was similar to the results from the previous experimental studies. After investigating the effects of TCDD on the liver at the histological level, the previous studies reported that parenchymal degeneration and the vacuolization of hepatocytes was observed in the rats that received TCDD [8] and degenerated hepatocytes, lobular inflammation and marked fat accumulation in TCDD-treated mice [25], although different doses of TCDD was used between the two studies. These results suggest that dioxin exposure causes alterations in liver morphology at the histological level, which was clearly exhibited by inflammation and degenerated hepatocytes.

Other indications of chronic liver disease, such as Mallory bodies, pigmentation, granulomatosis, cytoplasmic alterations, megamitochondria and venous obstruction, were also assessed but were not seen in any subjects in the present study. In a previous study, it was reported that alterations in cytoplasm, fatty change, bile duct hyperplasia and pigmentation were found in the livers of rats that were exposed to TCDD [26]. In a critical review of the histopathological findings of 118 papers on endocrine and non-endocrine hepatic toxicity in fish, Wolf et al., (2018) also reported that exposure to endocrine and non-endocrine hepatic toxicity, such as TCDD, induced cytoplasmic alterations, including Mallory bodies, pigmentation, granulomatosis, megamitochondria and venous obstruction [27]. However, these alterations were not detected in the present study. This could partly be explained by the expression of the effects of dioxin exposure on morphology at the histological level based the exposure levels, the length of exposure to dioxin and the location from which the tissues were collected [28]. In addition, the number of subjects in this study was small, which could also have affected results. Therefore, further studies with higher numbers of subjects are required to investigate the effects of dioxin exposure on histological changes in the liver.

Furthermore, in the present study, we found that increased exposure to TCDD and TEQ-PCDD/Fs was associated with increased METAVIR scores. We found that the most common score was F1, followed by F0 and F2. No cases of F3 and F4 scores were detected. The Vietnam Ministry of Health has classified primary liver cancer as on the list of diseases, malformations and congenital abnormalities that are associated with exposure to toxic chemicals/dioxin [29]. An increased prevalence of mortality from all causes of death, including liver cancer and liver cirrhosis, associated with exposure to dioxin that originated from the use of Agent Orange in Vietnam has also been reported [3–6]. Therefore, dioxin that originated from Agent Orange could potentially cause very serious liver damage, but no such serious liver damage was observed in the current study. It is possible that the above studies were mainly conducted on veterans, most of whom were directly exposed to the dioxin that was sprayed by the US during the war, and that these studies were conducted a relatively long time ago. Moreover, our study was conducted on people living in the dioxin contamination hot spot in Da Nang, who had longer periods of exposure to dioxin but at a lower concentration that decreased over time (due to the natural decomposition of dioxin and the dioxin detoxification process). Furthermore, the determination of the degree of liver fibrosis, i.e., significant or severe fibrosis, is suggested to be very important to establish the correct treatment method and to screen for complications. Most patients with significant fibrosis (≥F2) need immediate treatment to avoid progression to severe fibrosis. For severe fibrosis (≥F3), patients need special monitoring and screening for complications (liver cancer, gastrointestinal bleeding due to esophageal varices rupture, etc.). Therefore, regular health checks are necessary for people living in hot spots of dioxin contamination in Vietnam to establish the correct therapy to treat patients with F2 grade fibrosis and to prevent the progression of patients with F0 and F1 grade fibrosis.

The present study was the first study to report the effects of dioxin exposure that originated from Agent Orange on human health at the histological level in Vietnam. It is generally accepted that the toxicity of TCDD is induced mainly via the activation of the Ah receptor binding to the xenobiotic response element of target genes [30], leading to the induction or suppression of the transcription of numerous genes that have been linked to cancer development due to changes in tumor suppressor proteins, oncogenes, growth factors and cell cycle proteins, among other factors [31]. It has been suggested that TCDD promotes tumor progression in vivo by directly targeting mitochondrial transcription and the induction of mitochondrial stress signaling [32]. In addition, mitochondrial dysfunctions are frequently described as early and initiating events in various chronic pathological conditions in different tissues and organs, including the liver, brain and heart, and have been suggested as biomarkers for the early detection of cancer [33]. Alterations in the liver at the ultrastructural level, particularly in mitochondria, have been found in chronic hepatitis patients who were exposed to dioxin [10]. Therefore, we plan to investigate the associations between dioxin exposure and alterations in the liver at the ultrastructural level in the present subjects. Health regular checks, including checks for liver cancers and chronic diseases in other organs, is required in all subjects.

Limitations: In this study, we investigated alterations at the histological level in the livers of patients who were diagnosed with chronic hepatitis and were exposed to high levels of dioxin. However, our study had some limitations. We did not have a control group to compare the effects of dioxin exposure on the liver at the histological level. We initially selected 33 chronic hepatitis B patients living in unsprayed areas of Vietnam as the control group. However, there were only three male patients and no patients with F0 grade fibrosis and 22 patients (66.7%) showed higher than F1 grade fibrosis according to the METAVIR scale. Therefore, we decided that this group was not suitable as a control group for the present study. Other factors, such as food intake, were also not controlled in the present study, which could have affected the results of the laboratory tests. Another limitation of the present study was its relatively small sample size, particularly for the comparison of TCDD and TEQ-PCDD/Fs levels among different liver fibrosis grades.

5. Conclusions

This study showed alterations at the histological level in the livers of patients with chronic hepatitis who were exposed to dioxin that originated from Agent Orange. These alterations mainly comprised hydropic degenerative hepatocytes, lymphocytes and polynuclear leukocytes surrounding the liver cells and granular and lipoic degeneration. Furthermore, increased TCDD levels were associated with increased AST, ALT, protein and total bilirubin levels and liver fibrosis stage. Similarly, increased TEQ-PCDD/Fs levels were associated with higher levels of AST, protein and liver fibrosis stage. Regular health checks, including checks for dioxin-related diseases, should be required for all subjects living in dioxin contamination areas in Vietnam.

Author Contributions: Conceptualization, V.B.N. and P.Q.P.; methodology, V.B.N., N.X.D. and V.B.N.; investigation, P.Q.P., T.T.P., H.T.N., Q.V.H., T.D.N., T.M.H., D.T.D., Q.T.N.T., T.T.V. and T.M.L.; data curation, L.K.B., K.N., M.V.P., D.A.V. and T.N.P.; writing—original draft preparation, P.Q.P., T.N.P. and V.B.N.; writing—review and editing, T.N.P. and T.T.P.; funding acquisition, V.B.N. All authors have read and agreed to the published version of the manuscript.

Funding: This work was supported by the Vietnam Ministry of Science and Technology (KHCN-33.13/11-15). The funders had no role in the study design, the data collection and analysis, the decision to publish nor the preparation of the manuscript.

Institutional Review Board Statement: The study was conducted according to the guidelines of the Declaration of Helsinki and was approved by the Ethics Council of the Vietnam Military Medical University (Number: 33.13/11-15).

Informed Consent Statement: Informed consent was obtained from all subjects involved in the study.

Data Availability Statement: The data presented in this study are available on request to the corresponding author. The data are not publicly available due to the personal information (laboratory indices and the liver histology of the subjects).

Acknowledgments: We would like to thank all mothers and children who participated in this study. We would also like to thank the medical staff in the 17 military hospitals for their collaboration.

Conflicts of Interest: The authors declare no conflict of interest.

References

1. Schecter, A.; Dai, L.C.; Thuy, L.T.; Quynh, H.T.; Minh, D.Q.; Cau, H.D.; Phiet, P.H.; Nguyen, N.T.; Constable, J.D.; Baughman, R. Agent Orange and the Vietnamese: The persistence of elevated dioxin levels in human tissues. *Am. J. Public Health* **1995**, *85*, 516–522. [CrossRef]
2. Pocchiari, F.; Silano, V.; Zampieri, A. Human health effects from accidental release of tetrachlorodibenzo-p-dioxin (TCDD) at Seveso, Italy. *Ann. N. Y. Acad Sci.* **1979**, *320*, 311–320. [CrossRef]
3. Tamburro, C.H. Chronic liver injury in phenoxy herbicide-exposed Vietnam veterans. *Environ. Res.* **1992**, *59*, 175–188. [CrossRef]
4. O'Toole, B.I.; Marshall, R.P.; Grayson, D.A.; Schureck, R.J.; Dobson, M.; Ffrench, M.; Pulvertaft, B.; Meldrum, L.; Bolton, J.; Vennard, J. The Australian Vietnam Veterans Health Study: II. self-reported health of veterans compared with the Australian population. *Int. J. Epidemiol.* **1996**, *25*, 319–330. [CrossRef]
5. Kang, H.K.; Dalager, N.A.; Needham, L.L.; Patterson, D.G.; Lees, P.S., Jr.; Yates, K.; Matanoski, G.M. Health status of Army Chemical Corps Vietnam veterans who sprayed defoliant in Vietnam. *Am. J. Ind. Med.* **2006**, *49*, 875–884. [CrossRef] [PubMed]
6. Yi, S.W.; Ryu, S.Y.; Ohrr, H.; Hong, J.S. Agent Orange exposure and risk of death in Korean Vietnam veterans: Korean Veterans Health Study. *Int. J. Epidemiol.* **2014**, *43*, 1825–1834. [CrossRef] [PubMed]
7. Serdar, B.; LeBlanc, W.G.; Norris, J.M.; Dickinson, L.M. Potential effects of polychlorinated biphenyls (PCBs) and selected organochlorine pesticides (OCPs) on immune cells and blood biochemistry measures: A cross-sectional assessment of the NHANES 2003-2004 data. *Environ. Health* **2014**, *13*, 114. [CrossRef]

8. Czepiel, J.; Biesiada, G.; Gajda, M.; Szczepański, W.; Szypuła, K.; Dabrowski, Z.; Mach, T. The effect of TCDD dioxin on the rat liver in biochemical and histological assessment. *Folia Biol.* **2010**, *58*, 85–90. [CrossRef] [PubMed]
9. Fowler, B.A.; Lucier, G.W.; Brown, H.W.; McDaniel, O.S. Ultrastructural changes in rat liver cells following a single oral dose of TCDD. *Environ. Health Perspect.* **1973**, *5*, 141–148. [CrossRef] [PubMed]
10. Schecter, A.; Tiernan, T.; Schaffner, F.; Taylor, M.; Gitlitz, G.; VanNess, G.F.; Garrett, J.H.; Wagel, D.J. Patient fat biopsies for chemical analysis and liver biopsies for ultrastructural characterization after exposure to polychlorinated dioxins, furans and PCBs. *Environ. Health Perspect.* **1985**, *60*, 241–254. [CrossRef]
11. Dwernychuk, L.W. Dioxin hot spots in Vietnam. *Chemosphere* **2005**, *60*, 998–999. [CrossRef] [PubMed]
12. Hatfield Consultants, Office of National Committee 33, 2009. Comprehensive Assessment of Dioxin Contamination in Da Nang Airport, Viet Nam: Environmental Levels, Human Exposure and Options for Mitigating Impacts. (North Vancouver, BC, CANADA V7P0A3). Available online: http://www.hatfieldgroup.com/services/contaminant-monitoring-agent-orange/hatfield-agent-orange-reports-and-presentations/ (accessed on 21 January 2022).
13. Tai, P.T.; Nishijo, M.; Kido, T.; Nakagawa, H.; Maruzeni, S.; Naganuma, R.; Anh, N.T.; Morikawa, Y.; Luong, H.V.; Anh, T.H.; et al. Dioxin concentrations in breast milk of Vietnamese nursing mothers: A survey four decades after the herbicide spraying. *Environ. Sci. Technol.* **2011**, *45*, 6625–6632. [CrossRef]
14. Ministry of Health, Vietnam. *Decision Number 5448/ 5448/QĐ-BYT on the Guidance for Diagnosis, Treatment of Hepatitis B Disease (in Vietnamese: Quyết định số 5448/QĐ-BYT: Về việc ban hành hướng dẫn chẩn đoán, điều trị bệnh viêm gan vi rút B)*; Ministry of Health, Vietnam: Hanoi, Vietnam, 2014. Available online: http://kcb.vn/thu-vien-tai-lieu/huong-dan-chan-doan-dieu-tri-benh-viem-gan-vi-rut-b.html/ (accessed on 28 March 2022).
15. Olaf, P. PCDD/PCDF: Human background data for Germany, a 10-year experience. *Environ Health Perspect.* **1998**, *106*, 723–731.
16. Jules, L.D. Chronic Hepatitis. In *Harrison's Principles of Internal Medicine*, 19th ed.; McGraw Hill Education Medical: New York, NY, USA, 2015.
17. Maria, G.; Alessandra, M.; Gavino, F. Chronic viral hepatitis: The histology report. *Dig. Liver Dis.* **2011**, *43*, S331–S343.
18. Van den Berg, M.; Birnbaum, L.S.; Denison, M.; De Vito, M.; Farland, W.; Feeley, M.; Fiedler, H.; Hakansson, H.; Hanberg, A.; Haws, L.; et al. The 2005 World Health Organization reevaluation of human and mammalian toxic equivalency factors for dioxins and dioxin-like compounds. *Toxicol. Sci.* **2006**, *93*, 223–241. [CrossRef] [PubMed]
19. Poynard, T.; Ratziu, V.; Benmanov, Y.; Di Martino, V.; Bedossa, P.; Opolon, P. Fibrosis in Patients with Hepatitis C: Detection and Significance: Detection and Significance. *Semin. Liver Dis.* **2000**, *20*, 56–68. [CrossRef] [PubMed]
20. Van Luong, H.; Tai, P.T.; Nishijo, M.; Trung, D.M.; Thao, P.N.; Van Son, P.; Van Long, N.; Linh, N.T.; Nishijo, H. Association of dioxin exposure and reproductive hormone levels in men living near the Bien Hoa airbase, Vietnam. *Sci. Total Environ.* **2018**, *628–629*, 484–489. [CrossRef] [PubMed]
21. Triebig, G.; Werle, E.; Päpke, O.; Heim, G.; Broding, C.; Ludwig, H. Effects of dioxins and furans on liver enzymes, lipid parameters, and thyroid hormones in former thermal metal recycling workers. *Environ. Health Perspect.* **1998**, *106* (Suppl. S2), 697–700. [PubMed]
22. Calvert, G.M.; Hornung, R.W.; Sweeney, M.H.; Fingerhut, M.A.; Halperin, W.E. Hepatic and gastrointestinal effects in an occupational cohort exposed to 2,3,7,8-tetrachlorodibenzo-para-dioxin. *JAMA* **1992**, *267*, 2209–2214. [CrossRef] [PubMed]
23. Ott, M.G.; Zober, A.; Germann, C. Laboratory results for selected target organs in 138 individuals occupationally exposed to TCDD. *Chemosphere* **1994**, *29*, 2423–2437. [CrossRef]
24. Greig, J.B.; Jones, G.; Butler, W.H.; Barnes, J.M. Toxic effects of 2,3,7,8-tetrachlorodibenzo-p-dioxin. *Food Cosmet. Toxicol.* **1973**, *11*, 585–595. [CrossRef]
25. Lu, H.; Cui, W.; Klaassen, C.D. Nrf2 protects against 2,3,7,8-tetrachlorodibenzo-p-dioxin (TCDD)-induced oxidative injury and steatohepatitis. *Toxicol. Appl. Pharmacol.* **2011**, *256*, 122–135. [CrossRef]
26. Maronpot, R.R.; Foley, J.F.; Takahashi, K.; Goldsworthy, T.; Clark, G.; Tritscher, A.; Portier, C.; Lucier, G. Dose response for TCDD promotion of hepatocarcinogenesis in rats initiated with DEN: Histologic, biochemical, and cell proliferation endpoints. *Environ. Health Perspect.* **1993**, *101*, 634–642. [CrossRef]
27. Wolf, J.C.; Wheeler, J.R. A critical review of histopathological findings associated with endocrine and non-endocrine hepatic toxicity in fish models. *Aquat Toxicol.* **2018**, *197*, 60–78. [CrossRef]
28. Dobrzyński, M.; Madej, J.P.; Leśków, A.; Tarnowska, M.; Majda, J.; Szopa, M.; Gamian, A.; Kuropka, P. The Improvement of the Adaptation Process of Tocopherol and Acetylsalicylic Acid in Offspring of Mothers Exposed to TCDD. *Animals* **2021**, *11*, 3430. [CrossRef]
29. Ministry of Health, Vietnam. *The Decision Number 09/2008/QĐ-BYT: Promulgating the List of Diseases, Deformities, Deformities, RELATED TO EXPosure to Toxic Chemicals/Dioxins (in Vietnamsese Quyết định số 09/2008/QĐ-BYT: Ban hành Danh mục bệnh, tật, dị dạng, dị tật có liên quan đến phơi nhiễm với chất độc hóa học/dioxin)*; Ministry of Health, Vietnam: Hanoi, Vietnam, 2008. Available online: http://vbpl.yte.gov.vn/van-ban-phap-luat/quyet-dinh-092008qd-byt.1.1855.html/ (accessed on 28 March 2022).
30. Fernandez-Salguero, P.M.; Hilbert, D.M.; Rudikoff, S.; Ward, J.M.; Gonzalez, F.J. Arylhydrocarbon receptor-deficient mice are resistant to 2,3,7,8-tetrachlorodibenzo-pdioxin-induced toxicity. *Toxicol. Appl. Pharmacol.* **1996**, *140*, 173–179. [CrossRef]
31. Mead, M.N. Cancer and TCDD: The mitochondrial connection. *Environ. Health Perspect.* **2008**, *116*, A112.

32. Biswasm, G.; Srinivasan, S.; Anandatheerthavarada, H.K.; Avadhani, N.G. Dioxin-mediated tumor progression through activation of mitochondria-to-nucleus stress signaling. *Proc. Natl. Acad. Sci. USA* **2008**, *105*, 186–191. [CrossRef]
33. Modica-Napolitano, J.S.; Singh, K.K. Mitochondrial dysfunction in cancer. *Mitochondrion* **2004**, *4*, 755–762. [CrossRef]

toxics

MDPI

Article

Perinatal Dioxin Exposure and Attention Deficit Hyperactivity Disorder (ADHD) Symptoms in Children Living in a Dioxin Contamination Hotspot in Vietnam

Tai Pham-The [1,*], Muneko Nishijo [2], Thao Ngoc Pham [1], Hoa Thi Vu [2], Nghi Ngoc Tran [3], Anh Hai Tran [1], Luong Van Hoang [1], Quyet Do [1], Yoshikazu Nishino [2] and Hisao Nishijo [4]

[1] Institute of Biomedicine and Pharmacy, Vietnam Military Medical University, Hanoi 12108, Vietnam; phamngocthaovmmu@gmail.com (T.N.P.); anhhtr@yahoo.com (A.H.T.); hoangvanluong2003@yahoo.com (L.V.H.); dobaquyet@yahoo.com (Q.D.)
[2] Department of Public Health, Kanazawa Medical University, Uchinada, Ishikawa 920-0293, Japan; ni-koei@kanazawa-med.ac.jp (M.N.); vuhoa5593hvqy@gmail.com (H.T.V.); ynishi-no@kanazawa-med.ac.jp (Y.N.)
[3] Ministry of Health, Vietnamese Government, Hanoi 10060, Vietnam; nghi_tranngoc@yahoo.com
[4] Faculty of Medicine, University of Toyama, Toyama 930-0194, Japan; nishijo@med.u-toyama.ac.jp
* Correspondence: taithuy@kanazawa-med.ac.jp; Tel.: +84-339078560

Abstract: We examined children in Da Nang, a dioxin contamination hotspot in Vietnam, twice at 5 and 8 years of age, and investigated sex- and age-dependent differences in the effects of dioxin exposure on attention deficit hyperactivity disorder (ADHD) symptoms. We also studied autistic traits in children with ADHD symptoms. A total of 163 children participated in follow-up surveys at 5 and 8 years of age and were included in the present analysis. ADHD symptoms were assessed using an ADHD rating scale with inattention and hyperactivity-and-impulsivity (hyperactivity) subscales. Autistic behaviors were evaluated using the Autism Spectrum Rating Scale (ASRS). Perinatal dioxin exposure was indicated by dioxin levels in maternal breast milk. In boys, hyperactivity scores were significantly higher in the high 2,3,7,8-tetrachlorodibenzo-p-dioxin (TCDD) group only at 5 years of age. In girls, hyperactivity scores at 8 years of age were significantly higher in the high TCDD group, which was significantly associated with those at 5 years of age. In girls, ASRS unusual behavior scores were significantly higher with higher TCDD exposure and hyperactivity scores at 8 years of age. These results suggest that high perinatal TCDD exposure may increase ADHD likelihood and autistic traits, particularly in girls of 7–8 years of age.

Keywords: dioxins; birth cohort; attention deficit hyperactivity disorder (ADHD); autistic traits; sex difference; age difference

Citation: Pham-The, T.; Nishijo, M.; Pham, T.N.; Vu, H.T.; Tran, N.N.; Tran, A.H.; Hoang, L.V.; Do, Q.; Nishino, Y.; Nishijo, H. Perinatal Dioxin Exposure and Attention Deficit Hyperactivity Disorder (ADHD) Symptoms in Children Living in a Dioxin Contamination Hotspot in Vietnam. *Toxics* **2022**, *10*, 212. https://doi.org/10.3390/toxics10050212

Academic Editor: Michael Caudle

Received: 14 February 2022
Accepted: 1 April 2022
Published: 24 April 2022

1. Introduction

Endocrine disrupter chemicals (EDCs), such as polychlorinated biphenyls (PCBs), are suggested to have important roles in increasing the prevalence of neurodevelopmental disorders and behavior problems in children exposed to EDCs during the perinatal period [1]. Attention deficit hyperactivity disorder (ADHD) is one of the behavioral problems of which an increase has been of public concern in recent years [1]. Thus, associations between ADHD and EDCs, including PCBs [2–5], pesticides [6,7], and lead [8–10] exposure, have been investigated in contaminated areas in the United States and all over the world. However, there are not many previous studies investigating associations between ADHD symptoms and dioxins, which are among the most common EDCs [1], except for two birth cohort studies in dioxin-contaminated areas in Germany [11] and Italy [12].

From 2008 to 2009, we followed a birth cohort living in a dioxin-contaminated area in Da Nang, Vietnam, near a former U.S. military airbase, and investigated the effects of perinatal dioxin exposure on growth and neurodevelopment of children at various ages. Previously, we reported that lower planning ability associated with increasing toxic equivalency values of polychlorinated dibenzodioxins and polychlorinated dibenzofurans (TEQ-PCDD/Fs) in boys at 5 years of age [13]. In 8-year-old children, increased dioxin exposure was significantly associated with poor learning ability indicated by higher Colorado Learning Difficulties Questionnaire (CLDQ) scores, particularly in boys [14]. These results suggest that boys are susceptible to the effects of dioxin and their neurodevelopment was poorer than that of girls.

Regarding emotional and behavioral disorders, we reported increased children with autism spectrum disorder (ASD) behaviors associated with perinatal 2,3,7,8-tetrachlorodibenzo-p-dioxin (TCDD) exposure at 3 years of age for both sexes (boys are more likely than girls) [15]. In the survey of 8-year-olds, aggressive behavior was also examined by interviewing parents or caretakers of children using the Children's Scale of Hostility and Aggression: Reactive/Proactive (C-SHARP) with five subscales (verbal aggression, bullying, covert aggression, hostility, and physical aggression). The prevalence of high covert aggression scores in children of both sexes was significantly higher with higher TCDD exposure with an adjusted odds ratio of 4.5 [16]. However, in stratified analysis according to sex, the significant association between covert scores and TCDD exposure was clearer in girls than in boys [17]. When aggressive behavior is often observed in people with ADHD, it was suggestive and necessary to clarify the effects of dioxin on ADHD in this children cohort.

Furthermore, children with ASD often show inattention and impulsivity-hyperactivity and are diagnosed with ADHD as a comorbidity of ASD [18] because ASD and ADHD are reported to share etiological factors, such as inflammation in the brain [19,20] and disrupted gut microbiota that increases intestinal inflammation [21,22]. Hence, the aim of this present study was to investigate the associations between perinatal dioxin exposure and ADHD symptoms at different ages and sexes. In addition, the relationships between autistic behaviors and ADHD symptoms were analyzed to investigate increased autistic traits in children with increased ADHD symptoms.

2. Materials and Methods

2.1. Study Area and Subjects

Thanh Khe and Son Tra districts in Da Nang city were chosen as the study areas. These districts are located within 10 km of the former Da Nang airbase and are dioxin contamination hotspots originating from herbicide spraying in Vietnam. The subjects were children from the Da Nang birth cohort, including 238 mother-and-infant pairs (158 pairs in Thanh Khe and 80 pairs in Son Tra), recruited in 2008–2009 and followed up from 1 month to 8 years old [13–15,23–25].

A total of 168 mother–infant pairs (71.4% of birth cohort; 97 boys and 71 girls) participated in the follow-up surveys at 5 and 8 years of age. Because of missing behavioral examination data, 163 children (94 boys and 69 girls) were enrolled in the present analysis.

2.2. Perinatal Dioxin Exposure Levels Indicated by Dioxins in Breast Milk

The children's perinatal dioxin exposure was estimated using the dioxin levels in their mothers' breast milk collected a month after the birth of these children. Seventeen 2,3,7,8-substituted congeners of polychloro-dibenzodioxins (PCDDs) and polychloro-dibenzofurans (PCDFs) were measured using a high-resolution mass spectrometer (MStation-JMS700, JEOL, Tokyo, Japan), and the total toxic equivalents (TEQ) of PCDDs and PCDFs (TEQ-PCDD/Fs) were calculated by summing all of the values obtained by multiplying each congener concentration with reference to the WHO 2005 toxic equivalent factor [26]. The established method of analysis has been described previously in detail [27]. Because TCDD is the most toxic dioxin congener and specific for dioxin contamination originated from Agent Orange and TEQ-PCDD/Fs reflected the total toxic equivalent of all 17 PCDD/Fs, we selected these 2 indices as dioxin exposure markers in the present study.

2.3. Behavioral Assessment

The attention deficit hyperactivity disorder rating scale (ADHD-RS): parent ratings (2–5 years) was used to evaluate the children's behavioral disorders by parents or caretakers when children reached 5 and 8 years of age. The ADHD-RS includes two subscales, an inattention score (inattention) and impulsivity and hyperactivity (hyperactivity) score and a total scale score (ADHD). The values of each scale were standardized by age and sex with a reference to percentiles in the score sheets with the range of 1 to 99 percentiles [28]. Since this scale has no Vietnamese version, we translated this scale from the original version in English into Vietnamese and conducted a trial examination with 12 Vietnamese children to ensure the feasibility and appropriateness of this scale for the Vietnamese population.

Autistic behavior was evaluated using autism spectrum rating scales (ASRS) Autism Spectrum Rating Scales™ (ASRS®); Multi Health Systems Inc., (North Tonawanda, NY, USA) with three subscales, social communication (SC), unusual behavior (UB), and Diagnostic and Statistical Manual for Mental Disorders (DSM), and total scale (TOT) after percentile rank conversion from raw values to T-scores with the range of 25 to 85 using the technical manual for ASRS™ [29]. This scale has been translated from the English version into Vietnamese and applied to 179 3-year-old children from the Da Nang cohort [15].

Both the ADHD-RS and ASRS were standardized for children in the United States and not for those in Vietnam. Therefore, we cannot diagnose or estimate a risk of ADHD and ASRS for individual children based on the test results. However, a single examiner interviewed all participants for each scale to make scores reliable enough for indicating ADHD or ASD traits within group.

2.4. Statistical Analysis

SPSS (version 21.0) for Windows (IBM Corp., Armonk, NY, USA) was used to perform statistical analyses. TCDD concentrations and TEQs in breast milk were base-10 logarithmically transformed to normalize data distribution. Basing on the previous publications regarding the possible effective benchmarks of dioxin levels in this cohort, the cutoff values for high and low TCDD and TEQ-PCDD/Fs groups were set as 3.0 pg/g lipid and 17.6 pg-TEQ/g lipid, respectively. These cutoff values were derived from the 88th percentile concentration of TCDD and the 75th percentile level of TEQ-PCDD/Fs in breast milk [17,30].

General linear models were used to compare the mean ADHD-RS and ASRS scores between high and low TCDD or TEQ-PCDD/Fs groups after adjusting for covariates, including age, parity, maternal education, family income, gestational weeks, birth weight, and age in months at examination of children. Spearman's rho was used to analyze correlations between ADHD-RS scores at 5 and 8 years of age and between ADHD-RS scores and ASRS scores.

3. Results

3.1. The Characteristics of Mother-Child Pairs and Dioxin Exposure Levels

The characteristics of mothers and gestational weeks and weight at birth and age in months and body size at the time of the survey at 5 and 8 years old are shown in Table 1. There was no significant difference in the characteristics of mothers and children at birth between sexes. However, the mean BMI was significantly higher in boys at 5 and 8 years old compared with girls ($p < 0.05$ for t-test).

The geometric means with geometrical standard deviations of 17 PCDD/Fs congeners and TEQ-PCDD/Fs in maternal breast milk are shown according to child sex in Table 1. There was no significant difference of means for any congeners nor TEQ-PCDD/Fs between sexes. The mean TCDD of these samples was around three times higher than that in the unsprayed area (0.6 pg/g lipid) [27]. In addition, the mean TEQ-PCDD/Fs was four times higher than that in the unsprayed area (3.7 pg-TEQ/g lipid).

Table 1. Characteristics of the subjects and dioxin exposure in maternal breast milk.

Characteristics	Boys (N = 94) Mean, [N]	SD, (%)	Girls (N = 69) Mean, [N]	SD, (%)
Mothers				
Age (years)	28.2	5.9	28.8	6.6
Parity (% of primiparae)	[25]	(26.6)	[21]	(30.4)
Education (years)	8.7	3.3	8.4	3.6
Alcohol consumption during pregnancy (%)	[14]	(14.9)	[13]	(18.8)
Smoking habit of family members (%)	[82]	(87.2)	[59]	(85.5)
Family income (millions VNDs/month)	3.0	1.5	3.0	1.8
Children				
Gestational age (weeks)	39.6	0.8	39.6	0.8
Birth weight (g)	3259	395	3179	373
At the 5-year survey				
Age (months) at the survey	62.1	1.7	62.0	1.8
Weight (kg)	20.0	4.4	18.1	3.3
Height (cm)	109.1	4.5	107.9	4.1
BMI	16.7	2.6	15.5	2.1
At the 8-year survey				
Age (months) at the survey	92.8	1.4	93.1	1.5
Weight (kg)	28.0	7.1	24.2	4.9
Height (cm)	124.9	5.4	124.1	4.5
BMI	17.7	3.3	15.7	2.7
Dioxins in maternal breast milk *				
2,3,7,8-TCDD (pg/g-lipid)	1.4	2.0	1.5	2.6
1,2,3,7,8-PeCDD (pg/g-lipid)	4.3	1.6	4.3	1.7
1,2,3,4,7,8-HxCDD (pg/g-lipid)	2.3	1.6	2.4	1.6
1,2,3,6,7,8-HxCDD (pg/g-lipid)	8.4	1.6	8.4	1.8
1,2,3,7,8,9-HxCDD (pg/g-lipid)	2.7	1.6	2.7	1.7
1,2,3,4,6,7,8-HpCDD (pg/g-lipid)	12.1	1.5	12.5	1.6
OCDD (pg/g-lipid)	69.1	1.6	70.5	1.6
2,3,7,8-TCDF (pg/g-lipid)	0.5	1.9	0.5	2.2
1,2,3,7,8-PeCDF (pg/g-lipid)	1.2	1.7	1.3	2.0
2,3,4,7,8-PeCDF (pg/g-lipid)	7.2	1.5	7.5	1.7
1,2,3,4,7,8-HxCDF (pg/g-lipid)	17.0	1.7	18.9	1.8
1,2,3,6,7,8-HxCDF (pg/g-lipid)	10.5	1.7	11.7	1.8
1,2,3,7,8,9-HxCDF (pg/g-lipid)	0.2	2.2	0.3	2.6
2,3,4,6,7,8-HxCDF (pg/g-lipid)	1.2	1.6	1.4	1.8
1,2,3,4,6,7,8-HpCDF (pg/g-lipid)	11.7	1.7	13.4	1.9
1,2,3,4,7,8,9-HpCDF (pg/g-lipid)	1.2	2.0	1.2	2.6
OctaCDF (pg/g-lipid)	0.6	2.3	0.7	2.6
TEQ-PCDD/Fs (pg-TEQ/g-lipid)	12.9	1.5	13.4	1.7

N: number of subjects; SD: standard deviation; VNDs: Vietnamese Dongs; BMI: body mass index; *: Geometrical mean and geometrical standard.

3.2. Dioxin Exposure and ADHD-RS Scores at 5 and 8 Years of Age

For each sex, ADHD-RS scores for inattention, hyperactivity, and ADHD were examined at 5 years of age and compared between high and low TCDD and TEQ-PCDD/Fs exposure groups (Table 2). In boys, hyperactivity and ADHD scores were significantly higher in the high TCDD group ($\geq$3 pg/g lipid) than in the low TCDD group (<3). In girls, the hyperactivity scores and ADHD scores in the high TCDD group were more than 10 points higher than those in the low TCDD group; however, the differences between groups were not significant. For TEQ-PCDD/Fs, there was no significant difference in any ADHD-RS scores between the high and low exposure groups at the cut-off value of 17.6 (pg-TEQ/g lipid) in either sex (Table 2).

Table 2. Adjusted comparisons of ADHD scores at 5 years of age between high and low exposure groups.

ADHD-RS at 5 Years of Age	Low Exposure Mean	SD	Adj Mean	95% CI Lower	95% CI Upper	High Exposure Mean	SD	Adj Mean	95% CI Lower	95% CI Upper	p-Value
Boys											
TCDD	<3 (pg/g lipid) N = 85					≥3 (pg/g lipid) N = 9					
Inattention	30.9	28.1	30.4	24.7	36.1	46.3	26.4	50.4	31.4	69.4	0.052
Hyperactivity	36.8	31.3	36.2	29.9	42.6	54.2	27.7	59.8	38.7	80.8	0.039
ADHD	33.0	29.2	32.5	26.6	38.5	49.3	27.3	54.2	34.5	74.0	0.042
TEQ-PCDD/Fs	<17.6 (pg-TEQ/g lipid) N = 73					≥17.6 (pg-TEQ/g lipid) N = 21					
Inattention	32.3	27.6	32.0	25.6	38.5	32.5	31.1	33.4	20.6	46.2	0.859
Hyperactivity	39.3	30.9	40.4	33.3	47.5	35.8	33.2	31.8	17.6	45.9	0.301
ADHD	35.0	28.5	35.5	28.8	42.2	33.3	32.2	31.3	18.0	44.7	0.593
Girls											
TCDD	<3 (pg/g lipid) N = 57					≥3 (pg/g lipid) N = 12					
Inattention	42.0	27.6	42.1	34.7	49.6	51.4	27.9	50.6	33.4	67.8	0.382
Hyperactivity	41.5	32.1	41.0	32.2	49.8	49.8	37.7	52.1	31.8	72.4	0.331
ADHD	41.3	29.2	41.1	33.3	49.0	51.6	31.2	52.4	34.3	70.5	0.270
TEQ-PCDD/Fs	<17.6 (pg-TEQ/g lipid) N = 48					≥17.6 (pg-TEQ/g lipid) N = 21					
Inattention	42.4	28.5	44.3	35.9	52.7	46.3	25.9	42.1	28.8	55.4	0.795
Hyperactivity	43.4	33.0	44.8	34.9	54.7	42.0	34.0	38.7	23.1	54.4	0.537
ADHD	42.4	30.5	44.1	35.2	53.0	44.8	28.0	40.8	26.8	54.9	0.710

N: number of subjects; SD: standard deviation; adj mean: adjusted mean; CI: confidence interval; ADHD-RS: attention deficit hyperactivity disorder rating scale; Adjusted by age, parity, and education of mothers and gestational weeks, birth weight, month age at examination, and family income.

Three ADHD-RS scores, inattention, hyperactivity, and ADHD scores, examined in the 8-year-old survey were also compared between high and low TCDD and TEQ-PCDD/Fs groups, and the results for each sex are shown in Table 3. In boys, there was no significant difference in any ADHD-RS scale scores between high and low TCDD and TEQ-PCDD/Fs groups. Whereas, in girls, hyperactivity scores in high TCDD and TEQ-PCDD/Fs groups were significantly higher than those in low exposure groups, respectively.

3.3. Relationship between ADHD-RS Scores at 5 and 8 Years of Age

Higher ADHD-RS scores associated with increasing dioxin exposure were found in boys at 5 years of age and girls at 8 years of age, and therefore, Spearman's correlations between the ADHD-RS scores at 5 years and 8 years of age were analyzed in high and low TCDD groups and all children for each sex (Table 4). In the low exposure group and all children, almost all ADHD-RS scores at 8 years of age significantly correlated with those at 5 years of age for both sexes, except for the inattention scores in 8-year-old girls. However, in the high TCDD group, there was no correlation of any scale scores at 5 and 8 years of age in boys, while hyperactivity scores at 8 years of age significantly correlated with hyperactivity and ADHD scores at 5 years of age in girls.

3.4. Dioxin Exposure and ASRS Scale Scores at 5 Years of Age

In the survey at 5 years old, we also examined autistic traits in children using the ASRS scale and compared the mean SC, UB, DSM, and TOT scores between high and low dioxin exposure groups for each sex (Table 5). At this time, the ASRS scores for three boys and three girls were missing because there was no sufficient reply from these children's caretakers. The adjusted UB scores were significantly higher in the high TCDD group than in the low TCDD group only in girls. However, no significant difference in any ASRS scores between high and low TEQ-PCDD/Fs groups was observed in either girls or boys.

Table 3. Adjusted comparisons of ADHD scores at 8 years of age between high and low exposure groups.

ADHD-RS	Low Exposure					High Exposure					p-Value
	Mean	SD	Adj Mean	95% CI		Mean	SD	Adj Mean	95% CI		
at 8 Years of Age				Lower	Upper				Lower	Upper	
Boys											
TCDD			N = 85					N = 9			
Inattention	43.9	24.1	43.7	38.9	48.6	47.9	18.3	49.1	32.8	65.4	0.538
Hyperactivity	34.0	23.4	33.8	29.0	38.7	32.8	19.3	34.1	13.7	44.6	0.975
ADHD	37.5	21.5	37.3	33.1	41.5	39.4	14.7	40.7	26.6	54.9	0.650
TEQ-PCDD/Fs			N = 73					N = 21			
Inattention	45.2	23.9	44.2	38.8	49.5	41.1	22.8	44.5	33.8	55.1	0.965
Hyperactivity	36.2	22.9	36.2	31.0	41.5	25.5	21.7	25.5	15.1	35.9	0.081
ADHD	39.5	20.9	39.0	34.4	43.6	31.1	20.1	32.8	23.7	41.9	0.246
Girls											
TCDD			N = 57					N = 12			
Inattention	53.5	21.1	52.8	46.8	58.9	60.6	25.5	63.5	49.4	77.5	0.181
Hyperactivity	39.1	23.6	39.1	32.8	45.4	56.3	22.5	56.3	41.7	71.0	0.039
ADHD	44.7	21.9	44.3	38.1	50.6	57.6	26.1	59.5	45.0	73.9	0.065
TEQ-PCDD/Fs			N = 48					N = 21			
Inattention	54.2	21.5	53.9	47.0	60.7	55.9	23.1	56.6	45.7	67.4	0.691
Hyperactivity	37.0	22.0	37.2	30.3	44.2	53.8	25.2	53.2	42.2	64.2	0.023
ADHD	44.1	22.4	44.1	37.1	51.1	53.6	23.5	53.5	42.3	64.6	0.181

N: number of subjects; SD: standard deviation; adj mean: adjusted mean; CI: confidence interval; ADHD-RS: attention deficit hyperactivity disorder rating scale; Adjusted by age, parity, and education of mothers and gestational weeks, birth weight, month age at examination, and family income. Cut-off value for TCDD = 3.0 (pg/g lipid); Cut-off value for TEQ-PCDD/Fs = 17.6 (pg-TEQ/g lipid).

Table 4. Correlations between ADHD rating scores at 5 years and 8 years of age.

ADHD-RS	ADHD-RS at 5 Years of Age											
	Boys						Girls					
at 8 Years of Age	Inattention		Hyperactivity		ADHD		Inattention		Hyperactivity		ADHD	
TCDD < 3 (pg/g lipid)			N = 85						N = 57			
Inattention	0.457	***	0.391	***	0.451	***	0.218		0.195		0.210	
Hyperactivity	0.458	***	0.536	***	0.546	***	0.308	*	0.374	**	0.381	**
ADHD	0.539	***	0.562	***	0.598	***	0.296	*	0.327	*	0.334	*
TCDD ≥ 3 (pg/g lipid)			N = 9						N = 12			
Inattention	−0.172		−0.522		−0.263		0.004		0.359		0.151	
Hyperactivity	0.160		0.126		0.290		0.258		0.716	**	0.587	*
ADHD	−0.167		−0.413		−0.114		0.239		0.545		0.451	
All			N = 94						N = 69			
Inattention	0.422	***	0.339	***	0.409	***	0.220		0.262	*	0.262	*
Hyperactivity	0.434	***	0.498	***	0.515	***	0.368	**	0.449	***	0.462	***
ADHD	0.500	***	0.504	***	0.550	***	0.332	**	0.398	**	0.405	**

N: number of subjects; ADHD-RS: attention deficit hyperactivity disorder rating scale; *: $p < 0.05$, **: $p < 0.01$, ***: $p < 0.001$.

3.5. Relationship between ASRS and ASHD-RS Scores at 5 and 8 Years of Age

To clarify the relationship between the ADHD-RS and ASRS scores, the Spearman's correlation between ADHD scale and ASRS scale scores at 5 and 8 years of age was analyzed (Table 6). Significant correlations were found between all ADHD-RS scores at 5 years of age and UB scores in both sexes. In addition, in boys, inattention and ADHD scores also correlated with TOT scores. At 8 years of age, inattention and ADHD scores significantly correlated with TOT scores in boys. However, in 8-year-old girls, only the hyperactivity scores significantly correlated with UB scores.

Table 5. Adjusted comparisons of ASRS scores at 5 years of age between high and low exposure groups.

ASRS	Low Exposure					High Exposure					p-Value
	Mean	SD	Adj Mean	95% CI		Mean	SD	Adj Mean	95% CI		
at 5 Years of Age				Lower	Upper				Lower	Upper	
Boys											
TCDD			N = 83					N = 8			
SC	53.3	7.4	53.4	51.8	55.0	56.9	9.1	55.3	49.8	60.8	0.527
UB	56.6	7.1	56.5	55.1	57.9	58.3	8.4	58.7	53.7	63.6	0.416
DSM	56.4	4.9	56.5	55.4	57.6	60.1	7.4	59.1	55.3	62.8	0.205
TOT	55.4	4.5	55.5	54.5	56.5	58.1	6.3	57.6	54.1	61.0	0.255
TEQ-PCDD/Fs			N = 70					N = 21			
SC	53.3	7.6	53.2	51.4	55.0	54.6	7.9	55.0	51.5	58.5	0.381
UB	56.2	6.8	56.4	54.9	58.0	58.5	8.3	57.6	54.5	60.7	0.529
DSM	56.1	4.9	56.3	55.1	57.5	58.6	6.2	58.0	55.6	60.4	0.225
TOT	55.2	4.5	55.3	54.2	56.3	57.2	5.0	57.0	54.9	59.2	0.157
Girls											
TCDD			N = 55					N = 11			
SC	51.8	8.6	51.9	49.5	54.3	54.4	9.9	53.8	48.0	59.5	0.553
UB	57.2	5.0	57.2	55.8	58.6	61.3	6.2	61.3	57.9	64.6	0.031
DSM	55.1	5.4	55.1	53.5	56.7	57.9	8.7	57.7	53.9	61.6	0.229
TOT	54.9	4.8	55.0	53.5	56.4	58.7	7.3	58.2	54.8	61.7	0.089
TEQ-PCDD/Fs			N = 46					N = 20			
SC	53.4	9.1	53.5	50.8	56.1	49.4	7.3	49.1	44.8	53.5	0.110
UB	57.2	5.6	57.3	55.6	58.9	59.5	4.7	59.3	56.6	61.9	0.226
DSM	56.3	6.8	56.4	54.6	58.2	53.8	3.2	53.6	50.7	56.6	0.139
TOT	55.8	6.1	56.0	54.3	57.6	54.8	3.5	54.5	51.8	57.1	0.369

N: number of subjects; SD: standard deviation; adj mean: adjusted mean; CI: confidence interval; ASRS: autism spectrum rating scales; SC: Social communication; UB: Unusual behavior; DSM: Diagnostic and Statistical Manual for Mental Disorders; TOT: Total score; Adjusted by age, parity, and education of mothers and gestational weeks, birth weight, month age at examination, and family income. Cut-off value for TCDD = 3.0 (pg/g lipid); Cut-off value for TEQ-PCDD/Fs = 17.6 (pg-TEQ/g lipid).

Table 6. Correlations between ADHD rating scores at 5 and 8 years of age and ASRS scores at 5 years of age (Spearman's rho).

ADHD-RS	ASRS at 5 Years of Age			
	SC	UB	DSM	TOT
Boys (N = 91)				
At 5 years of age				
Inattention	−0.134	0.367 ***	0.143	0.290 **
Hyperactivity	−0.141	0.346 ***	0.067	0.172
ADHD	−0.133	0.375 ***	0.126	0.247 *
At 8 years of age				
Inattention	0.077	0.104	0.173	0.247 *
Hyperactivity	−0.085	0.164	0.045	0.125
ADHD	−0.031	0.196	0.145	0.235 *
Girls (N = 66)				
At 5 years of age				
Inattention	−0.176	0.337 **	0.031	0.008
Hyperactivity	−0.244 *	0.310 *	−0.081	−0.039
ADHD	−0.240	0.356 **	−0.045	0.006
At 8 years of age				
Inattention	0.123	−0.061	0.043	0.075
Hyperactivity	−0.216	0.358 **	−0.114	0.041
ADHD	−0.048	0.179	−0.014	0.098

N: number of subjects; ADHD-RS: attention deficit hyperactivity disorder rating scale; ASRS: autism spectrum rating scales; SC: Social communication; UB: Unusual behavior; DSM: Diagnostic and Statistical Manual for Mental Disorders; TOT: Total score; *: $p < 0.05$, **: $p < 0.01$, ***: $p < 0.001$.

4. Discussion

4.1. Perinatal Dioxin Exposure and ADHD Symptoms in Children Living in a Dioxin Contamination Hotspot in Vietnam

In the present study, increased ADHD symptoms were observed in children perinatally exposed to dioxins at higher levels than 3 (pg/g lipid) for TCDD and 17.6 (pg-TEQ/g lipid) for TEQ-PCDD/Fs compared with less exposed children. These exposure levels were similar to those which showed neurodevelopmental deficits in our previous studies for children from the Da Nang cohort [13–15]. However, dioxin effects on ADHD symptoms were clearer in girls than those in boys, in contrast with the studies of the same subjects (boys > girls) that targeted on other aspects of neurodevelopment such as motor coordination and higher cognitive development [13] and learning abilities in school [14].

In boys, hyperactivity and ADHD scores at 5 years of age were significantly higher in the high TCDD group than in the low TCDD group. However, we could not confirm the presence of increased children with ADHD traits based on these findings because 5 years old is not the eligible age for ADHD diagnosis (7–8 years old) with reference to the ADHD criteria in the DSM fourth edition [31].

In contrast, hyperactivity scores in girls at 8 years of age were significantly higher in the high TCDD and TEQ-PCDD/Fs groups than in the low exposure groups. In addition, in girls, hyperactivity scores at 8 years of age were significantly associated with those at 5 years of age, suggesting that ADHD symptoms, particularly impulsivity-hyperactivity, are increased at preschool age and reached levels sufficiently high to diagnose hyperactive-impulsive-type ADHD (ADHD-PH) in early school age (DSM fourth edition criteria). However, these results are inconsistent with two epidemiological facts in unexposed children with ADHD: (1) ADHD is more frequently found in boys and (2) a predominantly inattention type (ADHD-PI) is more common than ADHD-PH in girls [32]. These differences in ADHD characteristics between dioxin-exposed and non-exposed children might be related to the endocrine disrupter effects of dioxins, particularly TCDD. Moreover, the small number of subjects in high exposure group (n = 9 for boys) may decrease the statistical power in comparisons. Further large-scale studies from the perspective of sex differences are necessary in adolescents to clarify this issue.

In the present study, we analyzed child behavior based on a parent rating scale and did not examine their performance using neuropsychological function tests, such as continuous performance tasks (CPTs), which are commonly used to examine children with ADHD. Children with ADHD often have impaired neuropsychological functions, including vigilance, working memory, and response inhibition, which are related to their characteristic ADHD symptoms [32]. Therefore, in the future, CPTs should be performed in children to examine the neurobehavioral impairments associated with dioxin exposure and compare with those in non-exposed children with ADHD.

4.2. PCB and Dioxin Exposure and ADHD Symptoms in Children in Countries Other Than Vietnam

The WHO has carried out a series of exposure studies on the levels of PCDDs, PCDFs and PCBs in breast milk. The first WHO-coordinated exposure study took place in 1987–1988, the second round in 1992–1993, and the third round in 2000 to 2003 [33]. Temporal trends in the levels of PCDDs and PCDFs in human milk for countries around the world was observed continuously. For example, the decline between the levels found in the second round in 1993 and those found in the third round in 2003 is about 40%. In the third round, the lowest levels of PCDDs/PCDFs were found in countries in the Southern hemisphere (Fiji, Brazil, Philippines, Australia), from 3.34 to 5.57 pg-TEQ/g lipid, whereas the levels of PCDDs/PCDFs in European countries (Bulgaria, Croatia, Hungary, Ireland, Ukraine, Italy, Spain, Germany, Luxembourg) were comparatively high, from 6.14 to 14.97 pg-TEQ/g lipid. The countries in the group with the highest PCDD/Fs level included Belgium (16.92 pg-TEQ/g lipid), the Netherlands (18.27 pg-TEQ/g lipid), and Egypt (22.33 pg-TEQ/g lipid) [33,34]. In our present study, the level of TEQ- PCDDs/Fs was approximately 13 pg-TEQ/g lipid (Table 1), which is in the range of some European countries. The cutoff value for division into the low and high exposure groups was 17.6 pg-TEQ/g lipid, which was equal to the levels of countries in the highest group.

Associations between cord blood PCB and impulsivity examined using CPTs were reported at 4, 8, and 9 years of age in children in a birth cohort study in Oswego, USA [2,3]. A significantly increased risk of ADHD behavior indicated by Conner's teacher rating index in children with the highest quantile of PCB in cord blood was reported in children at 8 years of age whose mothers were residing in areas near a PCB-contaminated harbor in New Bedford, USA [4]. Sagiv et al. (2012) also examined these children using CPTs and showed significant associations between high rates of errors of omission and PCB exposure only in boys, suggesting sex-specific effects of PCBs on vigilance, an attention function [5].

In the Korean general population aged 12–15 years in the National Health and Nutrition Survey, a significantly increased prevalence of attention deficit disorder associated with increasing levels of serum dioxins, including HpCDD, OCDD, and HpCDF, was reported [35]; however, no rating scales or performance tests were used to estimate prevalence. In Germany, Neugebauer et al. (2015) examined attention in children aged 8–9 years from the Duisburg birth cohort exposed to dioxins and PCBs using a computer-based test battery KITAP and reported that significantly increased omission errors in divided attention were associated with increasing maternal blood levels of TEQ-PCBs and TEQ-PCDD/Fs [11]. However, they also reported conflicting results where parent rating ADHD scores were inversely associated with TEQ-PCBs levels in the same study subjects, suggesting no clear association between dioxin and PCB exposure and ADHD symptoms.

In children aged 7–17 years included in the Seveso second generation study in Italy from 2014 to 2016 whose mothers were exposed to extremely high levels of TCDD due to an industrial explosion in 1976, neurophysiological functions associated with attention and hyperactivity in children were examined using CPTs [12]. However, associations between prenatal TCDD exposure reflected by maternal serum TCDD collected in 1976 and ADHD index was found only in the children who were breastfed within 1 month [12], indicating no clear evidence of TCDD exposure on ADHD behavior.

In summary, these results from previous studies in countries other than Vietnam suggest perinatal EDCs exposure; for example, PCBs may increase children with ADHD symptoms at 8–9 years old [2–10]. However, the effects of perinatal dioxin exposure, particularly TCDD, which is the most toxic congener of dioxins [26], on ADHD symptoms and behavior were unclear in the dioxin studies [11,12] because there was no display of TCDD measurement values in the maternal samples during the perinatal period. We believe our birth cohorts in Vietnam are unique populations exposed to TCDD during perinatal period, which can provide more evidence for the effects of TCDD on neurophysiological function and behavior. Therefore, in the future, we need to conduct more studies with large populations with a variety of exposure levels in Vietnam to clarify the effects of TCDD on ADHD symptoms and behavior.

4.3. Associations between ASD and ADHD Symptoms

We previously reported increased autistic traits as reflected by ASRS scores in children with perinatal exposure of TCDD $\geq$ 3.5 (pg/g lipid) in both sexes (boys > girls) when children from the present birth cohort reached 3 years of age [15]. In the present study at 5 years of age, children with perinatal TCDD exposure $\geq$ 3.0 (pg/g lipid) also showed increased autistic traits. However, significantly increased scores were limited to the UB subscale and found only in girls. These differences in current ASRS examination results at 5 years of age from those at 3 years old may be partly caused by the lower cut-off value of the exposure group in the present study, meaning that the high TCDD group included moderately affected children, particularly with respect to boys. Moreover, some boys whose ASRS scores were high at 3 years of age might have acquired communication skills after the survey, resulting in no more typical autistic symptoms at 5 years of age. However, ADHD symptoms at 5 years of age significantly correlated with ASD-related unusual behaviors, suggesting that boys still have mildly increased autistic traits, which may increase ADHD-like behaviors at a preschool age.

In contrast, in girls, UB scores, which were significantly higher in the high TCDD group, significantly correlated with hyperactivity scores at 5 and 8 years of age. These results suggest that high TCDD exposure may increase behavioral disorders in the form of combined ASD and ADHD, which became clear at an early school age. For these children at 9 years of age, we also reported that TCDD exposure may influence the EEG power of the mirror neuron system of the brain, particularly in girls, which contributes to social-emotional behavior and is often found in children with ASD [30]. These neurophysiological findings support that girls exposed to high levels of TCDD may have atypical brain development, leading to combined ASD and ADHD, which has more notable symptoms than those of ASD alone at an early school age.

4.4. Associations between ADHD Symptoms, Learning Ability, and Aggressive Behavior

In boys aged 8 years, the Colorado Learning Difficulties Questionnaire (CLDQ) reading scores were significantly higher and reading errors of some passages in the high TCDD exposure group suggested learning difficulties at school [10]. However, the C-SHARP aggression examination also showed no increase in any type of aggression and there was no change in ADHD-RS scores in boys [17], suggesting boys with high TCDD exposure may show poor performance at school but had no active problematic behaviors, such as ADHD and aggressive behaviors.

In contrast, in girls, an increased covert aggression score was significantly associated with higher TCDD levels and ADHD-RS scores [17]. However, no difference in CLDQ scores and language achievement scores was found between high and low TCDD groups in girls at 8 years of age [14], suggesting no influence of dioxins on learning ability and school performance in girls. In our previous studies on neurodevelopment in children from the present Da Nang cohort from 4 months to 5 years old, the effect of dioxin exposure on neurodevelopmental indices, including cognitive, linguistic, and motor abilities including co-ordination motor skills, was found only in boys [13,14,23,24].

Anxiety and depression are common comorbidities in children with ADHD [32]. In future, we will follow-up these children into adolescence over 12 years of age to examine their mental health status using anxiety and depression scales, as well as the ADHD-RS and C-SHARP aggression scale, to clarify dioxin effects on social-emotional behavior in these children from Vietnamese birth cohorts.

4.5. Strength and Limitations

Our study reported increased children with ADHD associated with dioxin, particularly TCDD, exposure during the perinatal period. To our knowledge, this is the first study to report a significant association between perinatal TCDD exposure indicated by dioxin levels in maternal breast milk and ADHD. Similarly, previous studies in the USA reported the impact of prenatal PCB exposure on the increased prevalence of ADHD in children [2–4]; however, their exposure levels were assessed using PCB levels in the cord blood, which shows the levels to which the fetus was directly exposed. Despite this limitation of not assessing exposure levels, we have nevertheless shown good correlations between dioxin concentrations, particularly TCDD, in breast milk and cord blood [36], indicating that TCDD in breast milk can be a good exposure marker during the fetal period.

In addition, the sample size of high TCDD group, particularly in boys, is a limitation of the present study that decreases the statistical power in comparisons of ADHD-RS and ASRS scores between high and low exposure groups. In addition, we have no control group, which should be enrolled from unexposed populations in unsprayed north Vietnam. However, they have very different maternal education levels and economic statuses from those in Da Nang located in south Vietnam, which are too strong confounding factors for adjusting associations between dioxin exposure and increased children with ADHD.

Another limitation is the ADHD-RS, which is based on the ADHD criteria in the DSM fourth edition, which is an older version than the DSM fifth edition [37] used at present time. Furthermore, performance tests, such as CPTs, were not performed in the present study, and they should be used to obtain objective results in future.

5. Conclusions

Perinatal TCDD exposure temporally increased hyperactivity-and-impulsivity in boys only at a pre-school age (5 years). In contrast, in girls, TCDD exposure increased hyperactivity-and-impulsivity at an early school age (8 years) and unusual ASD behavior at a younger age. These results suggest that school-aged girls perinatally exposed to high TCDD levels are more likely to show ADHD symptoms combined with autistic traits than boys of the same age.

Author Contributions: Conceptualization, M.N., Y.N., H.N. and T.P.-T.; methodology, T.P.-T., T.N.P. and M.N.; investigation, T.P.-T., T.N.P., N.N.T. and M.N.; data curation, T.P.-T. and M.N.; writing—original draft preparation, H.T.V. and M.N.; writing—review and editing, T.P.-T. and M.N.; project administration, A.H.T., L.V.H. and Q.D.; funding acquisition, M.N. All authors have read and agreed to the published version of the manuscript.

Funding: This work was supported by the Japan Society for the Promotion of Science; Grant-in-Aid for Scientific Research (B) (17406016 and 17H04665).

Institutional Review Board Statement: The study was conducted in accordance with the Declaration of Helsinki and approved by the Institutional Review Board of the Health Department of Da Nang city and the Kanazawa Medical University with ethical clearance number No. I183 and approved on 27 July 2017.

Informed Consent Statement: Informed consent was obtained from all subjects involved in the study.

Data Availability Statement: The data presented in this study are available on request to the corresponding author. The data are not publicly available due to the personal information.

Acknowledgments: We thank all mother–infant pairs participating in our Da Nang birth cohort study. We are grateful to medical staffs in Da Nang city.

Conflicts of Interest: The authors declare no conflict of interest.

References

1. Silver, M.K.; Meeker, J.D. Chapter 14-Endocrine disruption of developmental pathways and children's health. In *Endocrine Disruption and Human Health*, 2nd ed.; Darbre, P.D., Ed.; Elsevier: Amsterdam, The Netherlands, 2021; pp. 291–320.
2. Stewart, P.; Fitzgerald, S.; Reihman, J.; Gump, B.; Lonky, E.; Darvill, T.; Pagano, J.; Hauser, P. Prenatal PCB exposure, the corpus callosum, and response inhibition. *Environ. Health Perspect.* **2003**, *111*, 1670–1677. [CrossRef] [PubMed]
3. Stewart, P.; Reihman, J.; Gump, B.; Lonky, E.; Darvill, T.; Pagano, J. Response inhibition at 8 and 9 1/2 years of age in children prenatally exposed to PCBs. *Neurotoxicol. Teratol.* **2005**, *27*, 771–780. [CrossRef]
4. Sagiv, S.K.; Thurston, S.W.; Bellinger, D.C.; Tolbert, P.E.; Altshul, L.M.; Korrick, S.A. Prenatal organochlorine exposure and behaviors associated with attention deficit hyperactivity disorder in school-aged children. *Am. J. Epidemiol.* **2010**, *171*, 593–601. [CrossRef] [PubMed]
5. Sagiv, S.K.; Thurston, S.W.; Bellinger, D.C.; Altshul, L.M.; Korrick, S.A. Neuropsychological measures of attention and impulse control among 8-year-old children exposed prenatally to organochlorines. *Environ. Health Perspect.* **2012**, *120*, 904–909. [CrossRef] [PubMed]
6. Bouchard, M.F.; Bellinger, D.C.; Wright, R.O.; Weisskopf, M.G. Attention-deficit/hyperactivity disorder and urinary metabolites of organophosphate pesticides. *Pediatrics* **2010**, *125*, e1270–e1277. [CrossRef] [PubMed]
7. Marks, A.R.; Harley, K.; Bradman, A.; Kogut, K.; Barr, D.B.; Johnson, C.; Calderon, N.; Eskenazi, B. Organophosphate pesticide exposure and attention in young Mexican-American children: The CHAMACOS study. *Environ. Health Perspect.* **2010**, *118*, 1768–1774. [CrossRef] [PubMed]

8. Nicolescu, R.; Petcu, C.; Cordeanu, A.; Fabritius, K.; Schlumpf, M.; Krebs, R.; Krämer, U.; Winneke, G. Environmental exposure to lead, but not other neurotoxic metals, relates to core elements of ADHD in Romanian children: Performance and questionnaire data. *Environ. Res.* **2010**, *10*, 476–483. [CrossRef] [PubMed]
9. Nigg, J.T.; Nikolas, M.; Knottnerus, G.M.; Cavanagh, K.; Friderici, K. Confirmation and extension of association of blood lead with attention-deficit/hyperactivity disorder (ADHD) and ADHD symptom domains at population-typical exposure levels. *J. Child Psychol. Psychiatry* **2010**, *51*, 58–65. [CrossRef]
10. Cho, S.C.; Kim, B.N.; Hong, Y.C.; Shin, M.S.; Yoo, H.J.; Kim, J.W.; Bhang, S.Y.; Cho, I.H.; Kim, H.W. Effect of environmental exposure to lead and tobacco smoke on inattentive and hyperactive symptoms and neurocognitive performance in children. *J. Child Psychol. Psychiatry* **2010**, *51*, 1050–1057. [CrossRef]
11. Neugebauer, J.; Wittsiepe, J.; Kasper-Sonnenberg, M.; Schoneck, N.; Scholmerich, A.; Wilhelm, M. The influence of low level pre- and perinatal exposure to PCDD/Fs, PCBs, and lead on attention performance and attention-related behavior among German school-aged children: Results from the Duisburg Birth Cohort Study. *Int. J. Hyg. Environ. Health* **2015**, *218*, 153–162. [CrossRef]
12. Ames, J.; Warner, M.; Siracusa, C.; Signorini, S.; Brambilla, P.; Mocarelli, P.; Eskenazi, B. Prenatal dioxin exposure and neuropsychological functioning in the Seveso Second Generation Health Study. *Int. J. Hyg. Environ. Health* **2019**, *222*, 425–433. [CrossRef] [PubMed]
13. Tran, N.N.; Pham, T.T.; Ozawa, K.; Nishijo, M.; Nguyen, A.T.; Tran, T.Q.; Hoang, L.V.; Tran, A.H.; Phan, V.H.; Nakai, A.; et al. Impacts of perinatal dioxin exposure on motor coordination and higher cognitive development in vietnamese preschool children: A five-year follow-up. *PLoS ONE* **2016**, *11*, e0147655. [CrossRef] [PubMed]
14. The, T.P.; Ngoc, T.P.; Van, T.H.; Nishijo, M.; Ngoc, N.T.; Thi, H.V.; Van, L.H.; Hai, A.T.; Nishino, Y.; Nishijo, H. Effects of perinatal dioxin exposure on learning abilities of 8-year-old children in Vietnam. *Int. J. Hyg. Environ. Health* **2020**, *223*, 132–141.
15. Nishijo, M.; Pham, T.T.; Nguyen, A.T.; Tran, N.N.; Nakagawa, H.; Hoang, L.V.; Tran, A.H.; Morikawa, Y.; Ho, M.D.; Kido, T.; et al. 2,3,7,8-Tetrachlorodibenzo-p-dioxin in breast milk increases autistic traits of 3-year-old children in Vietnam. *Mol. Psychiatry* **2014**, *19*, 1220–1226. [CrossRef]
16. Nishijo, M.; Pham, T.T.; Pham, N.T.; Duong, H.T.T.; Tran, N.N.; Kondoh, T.; Nishino, Y.; Nishimaru, H.; Do, Q.B.; Nishijo, H. Nutritional intervention with dried bonito broth for the amelioration of aggressive behaviors in children with prenatal exposure to dioxins in Vietnam: A pilot study. *Nutrients* **2021**, *13*, 1455. [CrossRef]
17. Pham-The, T.; Nishijo, M.; Pham-Ngoc, T.; Vu-Thi, H.; Tran-Ngoc, N.; Tran-Hai, A.; Hoang-Van, L.; Nishino, Y.; Nishijo, H. Effects of prenatal dioxin exposure on children behaviors at 8 years of age. In *Abstract Book, Proceedings of the 39th International Symposium on Halogenated Persistent Organic Pollutants-Dioxin in 2019, Kyoto, Japan, 20–30 August 2019*; Congress Secretariat: Osaka, Japan, 2019; p. 517.
18. Lord, C.; Elsabbagh, M.; Baird, G.; Veenstra-Vanderweele, J. Autism spectrum disorder. *Lancet* **2018**, *392*, 508–520. [CrossRef]
19. Theoharides, T.C.; Tsilioni, I.; Patel, A.B.; Doyle, R. Atopic diseases and inflammation of the brain in the pathogenesis of autism spectrum disorders. *Transl. Psychiatry* **2016**, *6*, e844. [CrossRef]
20. Dunn, G.A.; Nigg, J.T.; Sullivan, E.L. Neuroinflammation as a risk factor for attention deficit hyperactivity disorder. *Pharmacol. Biochem. Behav.* **2019**, *182*, 22–34. [CrossRef]
21. Petriello, M.C.; Hoffman, J.B.; Vsevolozhskaya, O.; Morris, A.J.; Hennig, B. Dioxin-like PCB 126 increases intestinal inflammation and disrupts gut microbiota and metabolic homeostasis. *Environ. Pollut.* **2018**, *242*, 1022–1032. [CrossRef]
22. Needham, B.D.; Kaddurah-Daouk, R.; Mazmanian, S.K. Gut microbial molecules in behavioural and neurodegenerative conditions. *Nat. Rev. Neurosci.* **2020**, *21*, 717–731. [CrossRef]
23. Tai, P.T.; Nishijo, M.; Anh, N.T.; Maruzeni, S.; Nakagawa, H.; Van Luong, H.; Anh, T.H.; Honda, R.; Kido, T.; Nishijo, H. Dioxin exposure in breast milk and infant neurodevelopment in Vietnam. *Occup. Environ. Med.* **2013**, *70*, 656–662. [CrossRef] [PubMed]
24. Tai, P.T.; Nishijo, M.; Nghi, T.N.; Nakagawa, H.; Van Luong, H.; Anh, T.H.; Nishijo, H. Effects of perinatal dioxin exposure on development of children during the first 3 years of life. *J. Pediatr.* **2016**, *175*, 159–166. [CrossRef] [PubMed]
25. Pham, T.T.; Nishijo, M.; Nguyen, A.T.N.; Tran, N.N.; Hoang, L.V.; Tran, A.H.; Nguyen, T.V.; Nishijo, H. Perinatal dioxin exposure and the neurodevelopment of Vietnamese toddlers at 1 year of age. *Sci. Total Environ.* **2015**, *536*, 575–581. [CrossRef] [PubMed]
26. Van den Berg, M.; Birnbaum, L.S.; Denison, M.; De Vito, M.; Farland, W.; Feeley, M.; Fiedler, H.; Hakansson, H.; Hanberg, A.; Haws, L.; et al. The 2005 World health organization reevaluation of human and mammalian toxic equivalency factors for dioxins and dioxin-like compounds. *Toxicol. Sci.* **2006**, *93*, 223–241. [CrossRef] [PubMed]
27. Tai, P.T.; Nishijo, M.; Kido, T.; Nakagawa, H.; Maruzeni, S.; Naganuma, R.; Anh, N.T.; Morikawa, Y.; Luong, H.V.; Anh, T.H.; et al. Dioxin concentrations in breast milk of Vietnamese nursing mothers: A survey four decades after the herbicide spraying. *Environ. Sci. Technol.* **2011**, *45*, 6625–6632. [CrossRef] [PubMed]
28. DuPaul, G.J.; Power, T.J.; Anastopoulos, A.D.; Reid, R. *ADHD Rating Scale—IV: Checklists, Norms, and Clinical Interpretation*; Guilford Press: New York, NY, USA, 1998.
29. Goldstein, S.; Naglieri, J.A. *Autism Spectrum Rating Scales™ (ASRS®)*; Multi-Health Systems Inc.: North Tonawanda, NY, USA, 2009.

30. Vu, H.T.; Nishijo, M.; Pham, T.N.; Pham-The, T.; Hoanh, L.V.; Tran, A.H.; Tran, N.N.; Nishino, Y.; Do, Q.; Nishijo, H. Effects of perinatal dioxin exposure on mirror neuron activity in 9-year-old children living in a hot spot of dioxin contamination in Vietnam. *Neuropsychologia* **2021**, *161*, 108001. [CrossRef] [PubMed]
31. American Psychiatric Association. *Diagnostic and Statistical Manual of Mental Disorders*, 4th ed.; text revision; American Psychiatric Association: Washington, DC, USA, 2000.
32. Aguiar, A.; Eubig, P.A.; Schantz, S.L. Attention deficit/hyperactivity disorder: A focused overview for children's environmental health researchers. *Environ. Health Perspect.* **2010**, *118*, 1646–1653. [CrossRef]
33. Van Leeuwen, F.X.R.; Malisch, R. Results of the third round of the who-coordinated exposure study on the levels of PCBs, PCDDs and PCDFs in human milk. *Organohalogen Compd.* **2002**, *56*, 311–316.
34. Malisch, R.; van Leeuwen, F.X.R. Results of the who-coordinated exposure study on the levels of PCBs, PCDDs and PCDFs in human milk. *Organohalogen Compd.* **2003**, *64*, 140–143.
35. Lee, D.H.; Jacobs, D.R.; Porta, M. Association of serum concentrations of persistent organic pollutants with the prevalence of learning disability and attention deficit disorder. *J. Epidemiol. Community Health* **2007**, *61*, 591–596. [CrossRef]
36. Boda, H.; Nghi, T.N.; Nishijo, M.; Thao, P.N.; Tai, P.T.; Van Luong, H.; Anh, T.H.; Morikawa, Y.; Nishino, Y.; Nishijo, H. Prenatal dioxin exposure estimated from dioxins in breast milk and sex hormone levels in umbilical cord blood in Vietnamese newborn infants. *Sci. Total Environ.* **2018**, *615*, 1312–1318. [CrossRef] [PubMed]
37. American Psychiatric Association. *Diagnostic and Statistical Manual of Mental Disorders*, 5th ed.; American Psychiatric Association: Arlington, VA, USA, 2013.

toxics

MDPI

Article

The Relationship of Dioxin Levels in Serum of 9-Year-Old Vietnamese Children and Their Mothers' Breast Milk

Ho Dung Manh [1], Teruhiko Kido [2,*], Takumi Takasuga [3], Michiko Yamashita [3], Le Minh Giang [4] and Hideaki Nakagawa [5]

[1] Faculty of Pharmacy, Lac Hong University, No. 10 Huynh Van Nghe, Buu Long, Bien Hoa 02513, Dong Nai, Vietnam; manhhodung@gmail.com
[2] Division of Health Sciences, Graduate School of Medical Science, Kanazawa University, 5-11-80 Kodatsuno, Kanazawa 920-0942, Japan
[3] Shimadzu Techno-Research Inc., 1 Nishinokyo Shimoaicho Nakagyo-ku, Kyoto 604-8436, Japan; t_takasuga00@shimadzu-techno.co.jp (T.T.); m_yamashita01@shimadzu-techno.co.jp (M.Y.)
[4] 10-80 Division, Hanoi Medical University, No. 1 Ton That Tung, Dong Da, Hanoi 116500, Vietnam; leminhgiang@hmu.edu.vn
[5] Department of Hygiene, Kanazawa Medical University, 1-1 Daigaku, Uchinada, Kahoku 920-0265, Japan; hnakagaw@kanazawa-med.ac.jp
* Correspondence: tkido@staff.kanazawa-u.ac.jp; Tel.: +81-762652565

Abstract: In this study, we measured the concentrations of polychlorinated dibenzo-p-dioxins (PCDDs) and polychlorinated dibenzofurans (PCDFs) in the blood of 9-year-old children living in a dioxin hotspot area and a nonexposed area in Vietnam. Forty-five blood samples were collected in the hotspot area while twelve pooled blood samples were collected in the nonexposed area. We found that the dioxin level of children in the hotspot was significantly higher than that of children in the nonexposed area. The total TEQ of PCDD/Fs in the hotspot and the nonexposed was 10.7 and 3.3 pg TEQ/g fat, respectively. However, TCDD, the maker of Agent Orange, was not detected in the blood of children in the hotspot area. In the hotspot area, four congeners 1,2,3,4,6,7,8-HpCDD, 1,2,3,4,7,8-HxCDF, 1,2,3,6,7,8-HxCDF, and 1,2,3,4,6,7,8-HpCDF in mothers' breast milk showed a significantly positive correlation with those in children's serum although the correlations of 1,2,3,7,8-PeCDD and 2,3,4,7,8-PeCDF were not significant. In addition, the duration of breastfeeding also correlates with dioxins in children. These results suggested that children in the hotspot area were exposed to dioxin through mothers' milk and other foods or environmental factors. The present study is the first study that shows dioxin levels in Vietnamese children.

Keywords: dioxin; Agent Orange; Vietnam; children

Citation: Manh, H.D.; Kido, T.; Takasuga, T.; Yamashita, M.; Giang, L.M.; Nakagawa, H. The Relationship of Dioxin Levels in Serum of 9-Year-Old Vietnamese Children and Their Mothers' Breast Milk. *Toxics* **2022**, *10*, 155. https://doi.org/10.3390/toxics10040155

Academic Editor: Kurunthachalam Kannan

Received: 4 February 2022
Accepted: 19 March 2022
Published: 25 March 2022

1. Introduction

During Operation Ranch Hand (1961−1971), the U.S. military sprayed millions of liters of herbicide Agent Orange over Southern Vietnam [1]. Unfortunately, the herbicide was contaminated with 2,3,7,8-tetrachlorodibenzo-p-dioxin (TCDD), which is one of the most toxic chemicals of the dioxins and is classified as a human carcinogen by the U.S. Environmental Protection Agency (EPA). More than 40 years have passed, and the dioxin levels are not very high in herbicide-sprayed areas due to degradation over a long time [2]. However, high dioxin levels have been still found in the soil inside former U.S. air bases as a result of herbicide spills, washing out herbicide tanks, and high-volume ground applications. These former air bases are now known as dioxin hotspots in Vietnam [3–5]. Recent studies on these hotspots have found that dioxin levels were very high in environmental and food samples [6]. In addition, the dioxin levels in breast milk and blood of Vietnamese peoples living around dioxin hotspot areas are also much higher than people living in nonexposed areas [2,7–15].

There is a growing concern about continuous dioxin exposure for Vietnamese people who live around dioxin hotspots. Recent studies have focused on the health effects of dioxin on children's development living around dioxin hotspots. These studies have shown that dioxin exposure has effects on endocrine disruption, physical development, and neurodevelopment in children [12,16–21]. However, these studies usually assess dioxin exposure in children by measuring dioxin levels in their mother's breast milk after the children are born. In this study, we will measure dioxin exposure in children by analyzing dioxin in their blood samples.

Several studies have shown that postnatal exposure to dioxins induces adverse effects on children. Studies in Chapaevsk, Russia have shown the association of peripubertal serum concentrations of organochlorines with growth, pubertal onset, and sexual maturity [22]. Studies in Seveso, Italy have shown that dioxin exposure in infancy/prepuberty lead to permanent effect on semen quality in human males as a result of the disruptive action of low concentrations of TCDD on the endocrine system [23].

To our knowledge, there is no study that has reported dioxin levels in children living in the hotspot areas in Vietnam. This may be due to difficulty in obtaining samples and the analytical technique. In this study, we aimed to measure and compare dioxin levels in blood of the 9-year-old children living in a dioxin hotspot area and a nonexposed area as well as assess the relationship between dioxin levels in children and their mother's breast milk.

2. Materials and Methods

2.1. Study Areas

The study areas are shown in Figure 1. Phu Cat Air Base is one of the three most dioxin-contaminated former U.S. air bases [5]. This air base is located in Phu Cat district, a rural area in Binh Dinh province. The TCDD level in the soil inside the air base has been found to be at a very high level of 236,000 pg/g taken at the herbicide storage area inside the air base [24], which suggests significant involvement of Agent Orange herbicide in soil/sediment samples because TCDD was the characteristic dioxin congener in Agent Orange. In the Phu Cat district, three subareas (PC1, PC2 and PC3) were selected according to their distance from the air base. The air base is inside PC1 (the Cat Tan and Ngo May communes, total area 40 km^2), near PC2 (Cat Tuong and Cat Trinh communes, total area 80 km^2), and a little farther from PC3 (Cat Hanh and Cat Lam communes, total area about 110 km^2).

The nonexposed area is Kim Bang district, which is located in Ha Nam province, in northern Vietnam. This area is also a rural area without an industrial zone nearby.

2.2. Study Participants

The initial study started in 2008. We collected breast milk from 60 mothers in the hotspot and 63 mothers from the nonexposed area when these children were 4–16 weeks [12]. In 2017, 45 blood samples of 9-year-old children were collected in the hotspot area and 35 blood samples of 9-year-old children were collected in the nonexposed area. The causes of loss in this follow-up study in both areas were due to traveling, visiting relatives, illness, or parent refusal. In the nonexposed area, due to the low level of dioxin, we combined each group of several samples to make a total of 12 pooled samples. In mother–child pair samples, we missed one breast milk sample from the hotspot and two breast milk samples from the nonexposed area. In this study, all mothers reported that their children were breastfed for 5 months on average (range from 1–9 months).

The purpose of the present study was thoroughly explained to them and written informed consent was obtained from each participant through their local people's committee. This study was approved by the Medical Ethics Committee of Kanazawa University (No. 455, 12 July 2013).

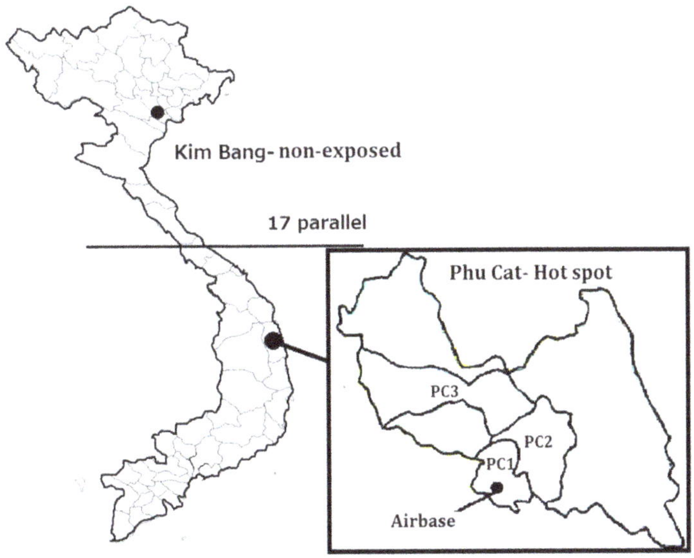

Figure 1. Study areas in Vietnam. The 17 parallel is the dividing line between North Vietnam and South Vietnam, as established by the 1954 Geneva Conference. PC1, PC2, and PC3 represent three subareas in the Phu Cat hotspot.

2.3. Dioxin Analysis

About 5 mL of blood were collected from each participant by medical staff at the community health centers. We then separated serum from samples and kept it frozen before sending it to Shimadzu Techno-Research Inc., in Kyoto, Japan for analysis. The analysis of 7 polychlorinated dibenzodioxins (PCDDs) and 10 polychlorinated dibenzofurans (PCDFs) in serum were described in the previous report [9]. After extraction and purification, the dioxins were quantified using high-resolution gas chromatography–high-resolution mass spectrometry (Hewlett-Packard 6890 Series and Micromass Autospec, Ultima). All PCDD/Fs congeners were calculated on a lipid basis which was determined from the crude extract by gravimetric method [25], then converted to toxic equivalents (TEQs) using the international toxicity equivalency factors (TEFs) 2005 recommended by the World Health Organization (WHO) [26]. The detection limits of congeners varied depending on the samples. Generally, a serum sample of 2.0 g wet weight with 0.5% lipid content (*w*/*w*) has detection limits, as follows: Te-PeCDD/Fs: 0.02 pg/g wet, 3 pg/g lipid; Hx-HpCDD/Fs: 0.04 pg/g wet, 6 pg/g lipid; OCDD/Fs: 0.1 pg/g wet, 10 pg/g lipid. Serum concentrations of dioxin and furan congeners below the limit of detection (LOD) were assigned a value equal to half the LOD.

Dioxin analysis in mothers' milk was described in our previous report [10]. Briefly, lipids were extracted from 10 g breast milk by liquid extraction and spiked with 40–80 pg of seventeen $^{13}C_{12}$-labeled PCDD/F congeners, as an internal standard. PCDD/Fs were purified on a multi-layer silica gel column and separated by an active carbon-dispersed silica gel column. The final sample extract was evaporated to dryness under a nitrogen stream then redissolved by addition of 20 μL of nonane containing 40 pg of $^{13}C_{12}$-1,2,3,4-TCDD and $^{13}C_{12}$-1,2,7,8-TCDF as external standards. PCDD/Fs were quantified using a gas chromatograph (HP-6980, Hewlett-Packard, Palo Alto, CA, USA) equipped with a high-resolution mass spectrometer (HRMS: JEOL MS station—JMS700). Analyses were performed in the selected ion-monitoring mode, and the resolution was maintained above 10,000. Seventeen PCDD/F congeners were calculated on a lipid basis and then converted to TEQ using the World Health Organization toxicity equivalency factors 2005. The detection limits were determined at a signal-to-noise ratio of 3. Samples with undetectable congener concentrations were assigned a value equal to half the LOD.

2.4. Data Analysis

We used R Statistical Environment (R Development Core Team, 2013) and Microsoft Excel 2010 (Microsoft Corp., Redmond, WA, USA) to conduct statistical analyses. Data are shown as geometric mean. The mean difference of each indicator between two areas or two genders was calculated by applying either Student's t-test or the Mann–Whitney U-test, depending on their distribution, as judged by a Shapiro–Wilk test. For comparison among three subareas in the hotspot, ANOVA test was used. Spearman's rank correlations were calculated between the concentrations of PCDD/F congeners in mothers' breast milk and the children's serum. The significance level was set to $p < 0.05$. Multiple linear regression was used to find the relationship of dioxin in the blood of children and dioxin in mother's breast milk, duration of breastfeeding, and children's gender. This information was obtained by interviewing the mothers during sample collection.

3. Results and Discussion

3.1. Serum Concentration of PCDD/Fs in Children

Table 1 shows dioxin levels in children in both areas. Only dioxin and furan congeners detected are shown. Four PCDD congeners, including 1,2,3,7,8-PeCDD, 1,2,3,6,7,8-HxCDD, 1,2,3,4,6,7,8-HpCDD, and OCDD are detected in both areas. The levels of these four PCDD congeners were also significantly higher in the hotspot than the nonexposed area. The other three PCDD congeners, including 2,3,7,8-TeCDD, 1,2,3,4,7,8-HxCDD, and 1,2,3,7,8,9-HxCDD, were below LOD in all samples.

Table 1. Dioxin levels in serum of children in the hotspot and the nonexposed areas.

pg/g Lipid	Hotspot (n = 45, Individual Samples)				Nonexposed (n = 12, Pooled Samples)				
	% over LOD	GM	GSD	Median	% over LOD	GM	GSD	Median	p-Value
PCDD congeners									
1,2,3,7,8-PeCDD	44	2.3	1.7	2.0	33	1.0	1.4	1.0	<0.0001 [a]
1,2,3,6,7,8-HxCDD	60	6.0	1.8	7.0	8	1.6	1.1	1.5	<0.0001 [a]
1,2,3,4,6,7,8-HpCDD	91	10.7	1.8	10.0	58	2.5	1.6	3.0	<0.0001 [a]
OCDD	100	161.6	1.6	160.0	100	64.3	1.3	63.0	<0.0001 [a]
PCDF congeners									
2,3,4,7,8-PeCDF	98	5.6	1.5	6.0	92	2.5	1.6	3.0	<0.0001 [a]
1,2,3,4,7,8-HxCDF	96	12.8	1.7	12.0	8	1.6	1.1	1.5	<0.0001 [a]
1,2,3,6,7,8-HxCDF	87	8.9	1.8	9.0	8	1.6	1.1	1.5	<0.0001 [a]
1,2,3,4,6,7,8-HpCDF	100	19.7	1.8	18.0	0	1.5	1.2	1.5	<0.0001 [a]
TEQ pg/g lipid									
TEQ PCDDs		5.5	1.4	5.3		1.8	1.3	1.5	<0.0001 [a]
TEQ PCDFs		5.1	1.4	4.7		1.6	1.3	1.6	<0.0001 [a]
TEQ PCDD/Fs		10.7	1.3	10.0		3.3	1.2	3.1	<0.0001 [a]

Note: n: number of subjects; GM: geometrical mean; GSD: geometrical standard deviation; LOD: limit of detection, [a] Mann–Whiney test.

Four PCDF congeners, including 2,3,4,7,8-PeCDF, 1,2,3,4,7,8-HxCDF, 1,2,3,6,7,8-HxCDF, and 1,2,3,4,6,7,8-HpCDF were detected in the hotspot area while only three congeners, 2,3,4,7,8-PeCDF, 1,2,3,4,7,8-HxCDF, and 1,2,3,6,7,8-HxCDF, were detected in lower percentage of samples in the nonexposed area. In particular, 1,2,3,4,6,7,8-HpCDF levels were detected in all samples in the hotspot while not detected in any sample in the nonexposed area. The levels of these PCDF congeners were also significant higher in the hotspot than the nonexposed area. Other PCDF congeners were not detected in both areas.

Table 2 shows dioxin levels in the hotspot area divided by gender and subareas. The proportion of males/females in the hotspot is 26/18, and in nonexposed is 13/22. This number is not similar in the two areas, but we did not find any difference in dioxin levels between males and females in the hotspot. In addition, there is no difference in dioxin levels among the children in the three subareas.

Table 2. Geometric mean of dioxin levels in hotspot divided by gender and three subareas.

pg/g Lipid	Gender		*p* Value [a]	Subareas			*p* Value [b]
	Male (n = 26)	Female (n = 18)		PC1 (n = 18)	PC2 (n = 15)	PC3 (n = 11)	
PCDD congeners							
1,2,3,7,8-PeCDD	2.2	2.6	0.2	2.3	2.2	2.6	0.9
1,2,3,6,7,8-HxCDD	5.9	6.1	0.7	5.7	5.3	7.4	0.2
1,2,3,4,6,7,8-HpCDD	11.0	10.0	0.4	11.2	9.8	10.8	0.8
OCDD	163.5	153.8	0.7	158.6	165.0	153.5	1.0
PCDF congeners							
2,3,4,7,8-PeCDF	5.9	5.1	0.4	5.4	5.3	6.5	0.4
1,2,3,4,7,8-HxCDF	13.4	11.7	0.3	11.8	11.5	16.3	0.1
1,2,3,6,7,8-HxCDF	10.0	7.5	0.1	8.6	7.6	11.9	0.1
1,2,3,4,6,7,8-HpCDF	19.7	19.6	0.7	18.0	20.1	22.0	0.3
TEQ pg/g lipid							
TEQ PCDDs	5.3	5.7	0.5	5.2	5.3	6.1	0.5
TEQ PCDFs	5.3	4.7	0.1	4.7	4.8	6.1	0.1
TEQ PCDD/Fs	10.7	10.6	1.0	10.0	10.3	12.3	0.1

Note: n: number of subjects; [a] Mann–Whiney test; [b] ANOVA test.

TCDD is the characteristic dioxin congener in Agent Orange. However, no sample has a TCDD level higher than its LOD. Previous studies have found TCDD in mothers' breast milk and blood of old men in the hotspot area around Phu Cat Air Base [9,10]. Furthermore, these studies also found that dioxin levels are significantly correlated with residential years. In this study, the children are only 9 years old. Besides dioxin exposure through breast milk, their exposure time to the external environment is short. In addition, when comparing dioxin levels in mothers' breast milk in three hotspots in Vietnam, we found that TCDD in breast milk in Phu Cat is low (range of 1–3 pg/g lipid) as compared with TCDD in breast milk samples collected in Bien Hoa (range of 0–27 pg/g lipid) or Da Nang (range of 1–10 pg/g lipid) hotspot [10]. One of the reasons is that people in Phu Cat live far from the air base, while in Bien Hoa city, people live near the air base. Therefore, it is necessary that we also biomonitor dioxin levels in children living in other hotspot areas such as Bien Hoa or Da Nang air base in future studies.

The children's serum concentration of total TEQ PCDD/Fs was 10.7 pg/g lipid in the hotspot, while it was 3.3 pg/g lipid in the nonexposed area. In Chapaevsk of Russia, an area with an extensive chemical manufacturing industry, the Khimprom chemical plant has produced chlorine-containing industrial and agricultural chemicals. The children of 8–9 years of age in this area have a median of TEQ PCDD/Fs level as high as 12.4 pg/g lipid in the blood samples collected between 2003–2005 [27]. In Kita borough of Tokyo, Japan where dioxin in the soil at the residential area is detected at a high level, the dioxin in the blood of 33 children (20 males and 13 females) whose age ranged from 7 to 15 years collected in 2006 is also high, with a mean of 5.5 pg/g lipid [28]. In German study, 10-year-old children living in industrial regions show a decreasing dioxin level. In 1993–1994, the average TEQ of PCDD/Fs was 10.2 pg/g lipid, and this level has decreased over time. In the most recent period of 2002–2003, the average of TEQ PCDD/Fs was reduced to 5.5 pg/g lipid [29]. There are a few studies worldwide that report dioxin levels in the blood of children, and therefore our study has added some valuable data on children's dioxin levels in the world.

3.2. *Correlation of Dioxin in Children's Blood and Mothers' Breast Milk*

Table 3 shows the correlation of dioxin in children and their mothers in the hotspot and the nonexposed area. In the hotspot area, four congeners, 1,2,3,4,6,7,8-HpCDD, 1,2,3,4,7,8-HxCDF, 1,2,3,6,7,8-HxCDF, and 1,2,3,4,6,7,8-HpCDF showed significant correlations between mothers' breast milk and children's serum. However, the correlations of 1,2,3,7,8-PeCDD, 2,3,4,7,8-PeCDF, and TEQ of PCDDs, PCDFs, or PCDD/Fs in mothers' milk and children's blood were not significant. This might be explained based on congeners' TEF (toxic equivalent factor), which is used to calculate TEQ. TEF of 1,2,3,7,8-PeCDD is 1 and is relatively high compared with those of HxCDD 0.1, HpCDD 0.01, and OCDD 0.0003 [26]. PCDFs also show a similar trend, with the highest TEF of 2,3,4,7,8-PeCDF being 0.3. This means that the PeCDD or PeCDF highly affect TEQ of PCDD, PCDF, and PCDD/Fs. Therefore, when the correlation of PeCDD and PeCDF was a very low value of −0.073 and 0.075, the correlations of PCDD, PCDF, and PCDD/Fs were also low and not significant. In addition, OCDD did not show a significant correlation in the hotspot. A previous study showed that OCDD levels are very high in food samples compared to other dioxin congeners [30]. The children are exposed to OCDD from both food intake and breast milk. Therefore, OCDD levels in children's blood did not correlate well with OCDD levels in their mother's milk in the hotspot.

Table 3. Spearman's rank correlation between the concentrations of PCDD/F congeners in mothers' breast milk and the children's serum.

	Hotspot (n = 44)		Nonexposed (Pooled Samples, n = 10)	
	r	*p*	r	*p*
PCDD congeners				
1,2,3,7,8-PeCDD	−0.073	0.637	0.461	0.180
1,2,3,6,7,8-HxCDD	0.159	0.303	−0.115	0.753
1,2,3,4,6,7,8-HpCDD	0.402	**0.007**	0.370	0.293
OCDD	0.140	0.365	0.675	**0.032**
PCDF congeners				
2,3,4,7,8-PeCDF	0.075	0.628	0.467	0.173
1,2,3,4,7,8-HxCDF	0.410	**0.006**	0.306	0.390
1,2,3,6,7,8-HxCDF	0.333	**0.027**	0.309	0.386
1,2,3,4,6,7,8-HpCDF	0.440	**0.003**	0.090	0.804
PCDDs TEQ	−0.047	0.763	0.663	**0.037**
PCDFs TEQ	0.237	0.121	0.642	**0.045**
PCDD/Fs TEQ	−0.028	0.854	0.705	**0.023**

In the nonexposed area, we have used pooled samples with an equal blood volume of several children. Therefore, we calculated the mean of dioxin levels in their mothers' breast milk samples. In the nonexposed area, most dioxin/furan congeners did not show a significant correlation between dioxin in mothers' milk and children's blood, but TEQ of PCDDs, PCDFs, and PCDD/Fs showed significant correlations. As we can see in Table 3, the correlation coefficients of most congeners were relatively high. For example, the correlation coefficients of 1,2,3,7,8-PeCDD and 2,3,4,7,8-PeCDF were 0.461 and 0.467, but their *p*-values were not significant. One reason might be due to the small number of samples (n = 10). However, because most individual congeners showed a similar trend of the high positive value of correlation coefficients, their TEQ values showed significant correlations even with a small number of samples.

Table 4 shows the relationship of dioxin level in children's serum and their gender, duration of full breastfeeding, and dioxin level in breast milk in the hotspot area assessed by using linear multiple regression. Gender shows no correlation with dioxin levels. Four dioxin congeners, 1,2,3,4,6,7,8-HpCDD, 1,2,3,4,7,8-HxCDF, 1,2,3,6,7,8-HxCDF, and 1,2,3,4,6,7,8-HpCDF, in breast milk have significant correlation with dioxin levels in children's serum. In addition, 2,3,4,7,8-PeCDF, 1,2,3,6,7,8-HxCDF, and PCDFs TEQ show correlation between dioxin in children's serum and duration of breastfeeding. In the nonexposed area, we could not assess such association due to pooled samples, not individual samples.

Table 4. Relationship of dioxin level in children's serum and their gender, duration of full breast feeding, and dioxin level in breast milk in the hotspot assessed by using linear multiple regression.

	Dioxin in Breast Milk (β)	Gender (β)	Duration of Breast Feeding (β)	R2
PCDD congeners				
1,2,3,7,8-PeCDD	−0.034	0.174	0.189	0.063
1,2,3,6,7,8-HxCDD	0.197	0.178	0.298	0.106
1,2,3,4,6,7,8-HpCDD	**0.321** *	0.054	0.209	0.143
OCDD	0.133	0.003	0.147	0.042
PCDF congeners				
2,3,4,7,8-PeCDF	0.035	0.102	**0.378** *	0.170
1,2,3,4,7,8-HxCDF	**0.460** **	0.115	0.261	0.226
1,2,3,6,7,8-HxCDF	**0.372** *	−0.007	**0.381** **	0.280
1,2,3,4,6,7,8-HpCDF	**0.426** **	0.173	0.257	0.209
TEQ PCDDs	−0.106	0.104	0.214	0.072
TEQ PCDFs	0.221	−0.006	**0.339** *	0.158
TEQ PCDD/Fs	−0.083	−0.021	0.278	0.086

Note: n = 44; β: standardized regression coefficient, * $p < 0.05$, ** $p < 0.01$.

The correlation of dioxin in children and their mothers suggested that one source of dioxin exposure in children is their mother's breast milk. In Vietnam, most children drink breast milk from when they are born. In this study, all mothers reported that their children were breastfed for 5 months on average (range from 1–9 months). The results are similar to the Russia study where 8- to 9-year-old boys who were breastfed for 26 weeks had a 28% increase in serum concentration compared with boys who were not breastfed [27]. In Kita borough, a dioxin-contaminated area in Tokyo, Japan, dioxin levels in 3- to 15-year-old children who were breastfed are about threefold higher than children that used formula milk. This study also found an association between the duration of breastfeeding and dioxin level in children [28]. In Germany, 10-year-old children living in industrial areas have 30% higher TEQ PCDD/Fs in the breastfeeding group than the non-breastfeeding group [29]. These studies only analyzed dioxin levels in children. However, our study measured dioxins in both children and mothers' breast milk and confirmed the dioxin exposure through mother's breastfeeding. Besides dioxin exposure through breast milk in their early stages, children are also exposed to dioxin through food intake. Food with high dioxin levels, such as free-range chicken meat and eggs, ducks, freshwater fish, snails, and beef, were reported in another hotspot area [31].

3.3. Comparison of Dioxin Congener Patterns in Children, Mothers, and Men in the Hotspot Area

We also compare the dioxin congener patterns in the serum of children and men and mothers' milk in the hotspot area. The congener pattern is expressed as a relative contribution to the sum of the measured concentrations. The serum samples of old men (mean age of 68 years old) who lived in this area were also collected in 2010–2011, and dioxin levels were described in the previous report [11]. Figure 2 shows a similar congener pattern between dioxin in children, mothers, and men. OCDD shows the highest contribution in men (62.1%), mothers (46.0%), and children (63.3%). Other PCDDs show the main contributions are 1,2,3,7,8-PeCDD, 1,2,3,6,7,8-HxCDD, 1,2,3,4,6,7,8-HpCDD, while four PCDFs show the major contributions are 2,3,4,7,8-PeCDF, 1,2,3,4,7,8-HxCDF, 1,2,3,6,7,8-HxCDF, and 1,2,3,4,6,7,8-HpCDF.

Recent dioxin decontamination projects were completed in Phu Cat Air Base where more than 7500 m^3 of dioxin-contaminated soil were contained in landfills in 2012 [32]. Therefore, the young generation may have little exposure to TCDD from the air base. Children are usually not exposed to occupational pollution and therefore represent a better trend of external exposure than adults. In the previous study, we found that men who live in PC1 (closest to the air base) have the highest dioxin levels; however, in this study, we do not find any difference in children in the three subareas. One reason might be that men also have occupational exposure, since in the previous study we found that TCDD was highest in two men (16 and 25 pg/g lipid) who had worked inside the air base [11]. Future studies on monitoring trend of dioxin levels in children will be necessary to reconfirm the effects of these remediation actions.

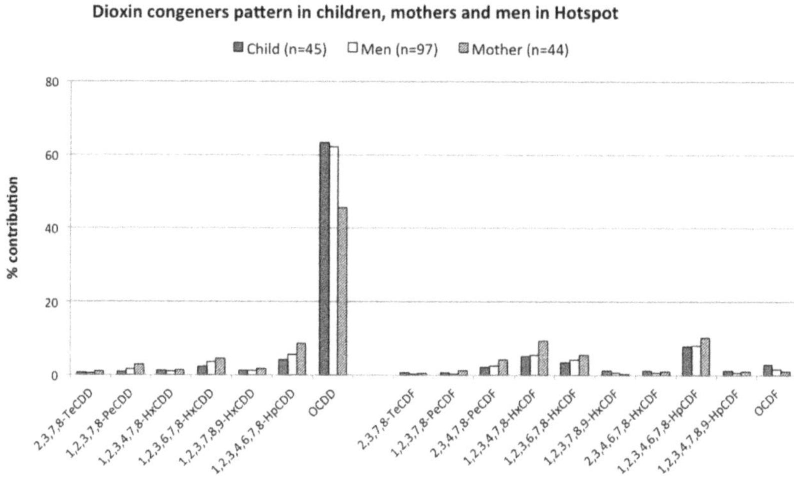

Figure 2. Congener patterns for PCDD/Fs in children, mothers, and men living in the hotspot area.

4. Conclusions

To our best knowledge, this is the first report showing dioxin levels in Vietnamese children living around a hotspot area. We found that dioxin levels of children living in the hotspot were significantly higher than those in the nonexposed area. The dioxin levels in children were correlated with dioxin level in mothers' breast milk and duration of breastfeeding. Future studies should focus on the health effects of dioxin on children and expand biomonitoring dioxin levels in children living in other hotspots in Vietnam.

Author Contributions: Conceptualization, T.K. and H.D.M.; methodology, T.K.; software, H.D.M.; validation, T.T. and M.Y.; formal analysis, T.T. and M.Y.; investigation, L.M.G.; resources, T.K.; data curation, H.D.M.; writing—original draft preparation, H.D.M.; writing—review and editing, T.K.; visualization, H.D.M.; supervision, H.N.; project administration, L.M.G.; funding acquisition, T.K. All authors have read and agreed to the published version of the manuscript.

Funding: This research was funded by a Grant-in-Aid for Scientific Research (B) from Japan Society for the Promotion of Science (20KK0220), and Lac Hong University (LHU-RF-MP-22-07-02).

Institutional Review Board Statement: The study was conducted in accordance with the Declaration of Helsinki, and approved by the Ethics Committee of Kanazawa University (No. 455). for studies involving humans.

Informed Consent Statement: Informed consent was obtained from all subjects involved in the study.

Data Availability Statement: The data presented in this study are available on request to the corresponding author.

Acknowledgments: The authors thank the study participants and the medical staff in the health centers of Phu Cat district and Kim Bang district.

Conflicts of Interest: The authors declare no conflict of interest. The funders had no role in the design of the study; in the collection, analyses, or interpretation of data; in the writing of the manuscript, or in the decision to publish the results.

References

1. Stellman, J.M.; Stellman, S.D.; Christian, R.; Weber, T.; Tomasallo, C. The extent and patterns of usage of Agent Orange and other herbicides in Vietnam. *Nature* **2003**, *422*, 681–687. [CrossRef]
2. Tawara, K.; Nishijo, M.; Maruzeni, S.; Nakagawa, H.; Kido, T.; Naganuma, R.; Suzuki, H.; Nhu, D.D.; Hung, N.N.; Thom, L.T.H. Residual congener pattern of dioxins in human breast milk in southern Vietnam. *Chemosphere* **2011**, *84*, 979–986. [CrossRef]
3. Dwernychuk, L.W. Dioxin hot spots in Vietnam. *Chemosphere* **2005**, *60*, 998–999. [CrossRef]
4. Dwernychuk, L.W.; Cau, H.D.; Hatfield, C.T.; Boivin, T.G.; Hung, T.M.; Dung, P.T.; Thai, N.D. Dioxin reservoirs in southern Viet Nam–a legacy of Agent Orange. *Chemosphere* **2002**, *47*, 117–137. [CrossRef]
5. Dwernychuk, L.; Hung, T.; Boivin, T. The Agent Orange dioxin issue in Vietnam: A manageable problem. *Organohalogen* **2006**, *68*, 312–315.
6. Le, L.T.H.; Dat, N.D.; Minh, N.H.; Nguyen, K.A. Characteristics of PCDD/Fs in soil and sediment samples collected from A-So former airbase in Central Vietnam. *Sci. Total Environ.* **2019**, *661*, 27–34. [CrossRef]
7. Hue, N.T.M.; Nam, V.D.; Thuong, N.V.; Huyen, N.T.; Phuong, N.T.H.; Hung, N.X.; Tuan, N.H.; Son, L.K.; Minh, N.H. Determination of PCDD/Fs in breast milk of women living in the vicinities of Da Nang Agent Orange hot spot (Vietnam) and estimation of the infant's daily intake. *Sci. Total Environ.* **2014**, *491-492*, 212–218. [CrossRef]
8. Hue, N.T.M.; Van Thuong, N.; Mai, P.T.N.; Minh, N.H. Site-specific bioaccumulation of polychlorinated dibenzo-p-dioxins and polychlorinated dibenzofurans (PCDD/PCDFs) in mothers and their infants living in vicinity of Bien Hoa airbase, Southern Vietnam. *Environ. Geochem. Health* **2018**, *40*, 2539–2549. [CrossRef]
9. Manh, H.D.; Kido, T.; Naganuma, R.; Xian Liang, S.; Supratman, S.; Anh, L.T.; Maruzeni, S.; Nishijo, M.; Nakagawa, H.; Honma, S.; et al. Serum Dioxin Levels in Vietnamese Men more than 40 Years after Herbicide Spraying. *Environ. Sci. Technol.* **2014**, *48*, 3496–3503. [CrossRef]
10. Manh, H.D.; Kido, T.; Tai, P.T.; Okamoto, R.; Honma, S.; Liang, S.X.; Anh, L.T.; Maruzeni, S.; Nghi, T.N.; Nishijo, M.; et al. Levels of polychlorinated dibenzodioxins and polychlorinated dibenzofurans in breast milk samples from three dioxin-contaminated hotspots of Vietnam. *Sci. Total Environ.* **2015**, *511*, 416–422. [CrossRef]
11. Nghi, T.N.; Nishijo, M.; Manh, H.D.; Tai, P.T.; Van Luong, H.; Anh, T.H.; Thao, P.N.; Trung, N.V.; Waseda, T.; Nakagawa, H.; et al. Dioxins and nonortho PCBs in breast milk of vietnamese mothers living in the largest hot spot of dioxin contamination. *Environ. Sci. Technol.* **2015**, *49*, 5732–5742. [CrossRef]
12. Oyama, Y.; Phuc, H.D.; Honma, S.; Oanh, N.T.P.; Hung, N.X.; Anh, L.T.; Manh, H.D.; Van Tung, D.; Nhu, D.D.; Tan, N.M.; et al. Decreased serum testosterone levels associated with 17β-hydroxysteroid dehydrogenase activity in 7-year-old children from a dioxin-exposed area of Vietnam. *Sci. Total Environ.* **2021**, *783*, 146701. [CrossRef]
13. Pham, D.T.; Nguyen, H.M.; Boivin, T.G.; Zajacova, A.; Huzurbazar, S.V.; Bergman, H.L. Predictors for dioxin accumulation in residents living in Da Nang and Bien Hoa, Vietnam, many years after Agent Orange use. *Chemosphere* **2015**, *118*, 277–283. [CrossRef]
14. Tai, P.T.; Nishijo, M.; Kido, T.; Nakagawa, H.; Maruzeni, S.; Naganuma, R.; Anh, N.T.N.; Morikawa, Y.; Van Luong, H.; Anh, T.H.; et al. Dioxin concentrations in breast milk of Vietnamese nursing mothers: A survey four decades after the herbicide spraying. *Environ. Sci. Technol.* **2011**, *45*, 6625–6632. [CrossRef]
15. Van Manh, P.; Tai, P.T.; Phuong, N.M.; Nishijo, M.; Trung, D.M.; Thao, P.N.; Son, H.A.; Van Tuan, T.; Van Chuyen, N.; Van Long, N.; et al. Serum dioxin concentrations in military workers at three dioxin-contaminated airbases in Vietnam. *Chemosphere* **2021**, *266*, 129024. [CrossRef]
16. Anh, L.T.; Kido, T.; Honma, S.; Manh, H.D.; Koike, I.; Oyama, Y.; Phuc, H.D.; Okamoto, R.; Nakagawa, H.; Nakayama, S.F.; et al. A relationship in adrenal androgen levels between mothers and their children from a dioxin-exposed region in Vietnam. *Sci. Total Environ.* **2017**, *607–608*, 32–41. [CrossRef]
17. Kido, T.; Honma, S.; Nhu, D.D.; Manh, H.D.; Van Tung, D.; Liang, S.X.; Anh, L.T.; Okamoto, R.; Maruzeni, S.; Nakagawa, H.; et al. Inverse association of highly chlorinated dioxin congeners in maternal breast milk with dehydroepiandrosterone levels in three-year-old Vietnamese children. *Sci. Total Environ.* **2016**, *550*, 248–255. [CrossRef]
18. Nishijo, M.; Pham, T.T.; Nguyen, A.T.N.; Tran, N.N.; Nakagawa, H.; Hoang, L.V.; Tran, A.H.; Morikawa, Y.; Ho, M.D.; Kido, T.; et al. 2,3,7,8-Tetrachlorodibenzo-p-dioxin in breast milk increases autistic traits of 3-year-old children in Vietnam. *Mol. Psychiatry* **2014**, *19*, 1220–1226. [CrossRef]
19. Oanh, N.T.P.; Kido, T.; Honma, S.; Oyama, Y.; Anh, L.T.; Phuc, H.D.; Viet, N.H.; Manh, H.D.; Okamoto, R.; Nakagawa, H.; et al. Androgen disruption by dioxin exposure in 5-year-old Vietnamese children: Decrease in serum testosterone level. *Sci. Total Environ.* **2018**, *640–641*, 466–474. [CrossRef]
20. Tai, P.T.; Nishijo, M.; Nghi, T.N.; Nakagawa, H.; Van Luong, H.; Anh, T.H.; Nishijo, H. Effects of Perinatal Dioxin Exposure on Development of Children during the First 3 Years of Life. *J. Pediatr.* **2016**, *175*, 159–166.e2. [CrossRef]
21. Van Tung, D.; Kido, T.; Honma, S.; Manh, H.D.; Nhu, D.D.; Okamoto, R.; Maruzeni, S.; Nishijo, M.; Nakagawa, H.; Ngoc, P.T.; et al. Low birth weight of Vietnamese infants is related to their mother's dioxin and glucocorticoid levels. *Environ. Sci. Pollut. Res.* **2016**, *23*, 10922–10929. [CrossRef] [PubMed]
22. Sergeyev, O.; Burns, J.S.; Williams, P.L.; Korrick, S.A.; Lee, M.M.; Revich, B.; Hauser, R. The association of peripubertal serum concentrations of organochlorine chemicals and blood lead with growth and pubertal development in a longitudinal cohort of boys: A review of published results from the Russian Children's Study. *Rev. Environ. Health* **2017**, *32*, 83–92. [CrossRef] [PubMed]

23. Mocarelli, P.; Gerthoux, P.M.; Patterson, D.G.; Milani, S.; Limonta, G.; Bertona, M.; Signorini, S.; Tramacere, P.; Colombo, L.; Crespi, C.; et al. Dioxin exposure, from infancy through puberty, produces endocrine disruption and affects human semen quality. *Environ. Health Perspect.* **2008**, *116*, 70–77. [CrossRef] [PubMed]
24. Vietnam-Russia Tropical Center and Hatfield Consultants. Evaluation of Contamination At the Agent Orange Dioxin Hot Spots in Bien Hoa, Phu Cat and Vicinity, Viet Nam. 2009. Available online: https://www.hatfieldgroup.com/wp-content/uploads/2020/05/undp1391-final-vrt-report-and-appendices.pdf (accessed on 20 February 2022).
25. Takasuga, T.; Senthilkumar, K.; Takemori, H.; Ohi, E.; Tsuji, H.; Nagayama, J. Impact of FEBRA (fermented brown rice with *Aspergillus oryzae)* intake and concentrations of PCDDs, PCDFs and PCBs in blood of humans from Japan. *Chemosphere* **2004**, *57*, 1409–1426. [CrossRef]
26. Van den Berg, M.; Birnbaum, L.S.; Denison, M.; De Vito, M.; Farland, W.; Feeley, M.; Fiedler, H.; Hakansson, H.; Hanberg, A.; Haws, L.; et al. The 2005 World Health Organization reevaluation of human and mammalian toxic equivalency factors for dioxins and dioxin-like compounds. *Toxicol. Sci.* **2006**, *93*, 223–241. [CrossRef]
27. Burns, J.S.; Williams, P.L.; Sergeyev, O.; Korrick, S.; Lee, M.M.; Revich, B.; Altshul, L.; Patterson, D.G.; Turner, W.E.; Needham, L.L.; et al. Predictors of serum dioxins and PCBs among peripubertal Russian boys. *Environ. Health Perspect.* **2009**, *117*, 1593–1599. [CrossRef]
28. Tohyama, C.; Uchiyama, I.; Hoshi, S.; Hijia, M.; Miyata, H.; Nagai, M.; Nakai, S.; Yauchi, M.; Ohkubo, S. Polychlorinated dioxins, furans, and biphenyls in blood of children and adults living in a dioxin-contaminated area in Tokyo. *Environ. Health Prev. Med.* **2011**, *16*, 6–15. [CrossRef]
29. Link, B.; Gabrio, T.; Zoellner, I.; Piechotowski, I.; Paepke, O.; Herrmann, T.; Felder-Kennel, A.; Maisner, V.; Schick, K.H.; Schrimpf, M.; et al. Biomonitoring of persistent organochlorine pesticides, PCDD/PCDFs and dioxin-like PCBs in blood of children from South West Germany (Baden-Wuerttemberg) from 1993 to 2003. *Chemosphere* **2005**, *58*, 1185–1201. [CrossRef]
30. Minh, N.H.; Tran, T.M.; Hue, N.T.M.; Minh, T.B.; Tuyet-Hanh, T.T. Bioaccumulation of PCDD/Fs in foodstuffs near Bien Hoa and Da Nang airbases: Assessment on sources and distribution. *Environ. Sci. Pollut. Res.* **2019**, *26*, 28852–28859. [CrossRef]
31. Tuyet-Hanh, T.T.; Minh, N.H.; Vu-Anh, L.; Dunne, M.; Toms, L.M.; Tenkate, T.; Thi, M.H.N.; Harden, F. Environmental health risk assessment of dioxin in foods at the two most severe dioxin hot spots in Vietnam. *Int. J. Hyg. Environ. Health* **2015**, *218*, 471–478. [CrossRef]
32. Boivin, T.; Hang, T.M. Evaluation of Dioxin Project Impact To Environment and People Evaluation of Dioxin Project Impact on the Environment and People. 2015. Available online: https://www.vn.undp.org/content/vietnam/en/home/library/environment_climate/evaluation_of_dioxin_project_impact_to_environment_and_people.html (accessed on 22 February 2022).

toxics

MDPI

Article

Effect of Perinatal Dioxin Exposure Originating from Agent Orange on Gaze Behavior in 3-Year-Old Children Living in the Most Dioxin-Contaminated Areas in Vietnam

Thao Ngoc Pham [1,2], Muneko Nishijo [1,*], Tai Pham-The [3], Nghi Ngoc Tran [4], Hoa Thi Vu [1], Anh Hai Tran [3], Tien Viet Tran [2], Yoshikazu Nishino [1] and Hisao Nishijo [5]

1 Department of Epidemiology and Public Health, Kanazawa Medical University, Ishikawa 920-0293, Japan; phamngocthaovmmu@gmail.com (T.N.P.); vuhoa5593hvqy@gmail.com (H.T.V.); ynishino@kanazawa-med.ac.jp (Y.N.)
2 Department of Functional Diagnosis, 103 Military Hospital, Vietnam Military Medical University, Hanoi 12108, Vietnam; tientv@vmmu.edu.vn
3 Biomedical and Pharmaceutical Research Centre, Vietnamese Military Medical University, Hanoi 12108, Vietnam; taithuy@kanazawa-med.ac.jp (T.P.-T.); anhhtr@yahoo.com (A.H.T.)
4 Ministry of Health, Vietnamese Government, Hanoi 10060, Vietnam; nghi_tranngoc@yahoo.com
5 Faculty of Medicine, University of Toyama, Toyama 930-0194, Japan; nishijo@med.u-toyama.ac.jp
* Correspondence: ni-koei@kanazawa-med.ac.jp; Tel.: +81-762188093

Abstract: We investigated the effect of perinatal dioxin exposure indicated by dioxins in breast milk on children's gaze behavior. We studied 142 children aged 3 years from the 2012 Bien Hoa birth cohort in a hotspot of dioxin contamination in Vietnam. Children's faces were viewed using the eye-tracking method. Associations between gaze behavior of faces and neurodevelopmental indices and head circumference were analyzed to determine whether poor gaze behavior indicates increased autistic traits in these children. The gaze fixation duration on facial areas when viewing 10 still images of children was calculated as the gaze behavior index. Autistic behavior was assessed using the Autism Spectrum Rating Scale, and language development was evaluated by the Bayley Scales of Infant and Toddler Development, Ver. 3. The face fixation duration (%) significantly decreased as 2,3,7,8-tetrachlorodibenzo-p-dioxin (TCDD) concentrations increased in a dose–effect manner in girls, which suggested atypical gaze behavior for watching human faces. Furthermore, these girls with atypical gaze behavior showed lower social communication scores and smaller head sizes, suggesting increased autistic traits in girls. In conclusion, our findings show sex-specific effects (girls > boys) of perinatal TCDD exposure on gaze behavior in young children.

Keywords: dioxin; gaze behavior; autistic trait; social communication; Vietnam

Citation: Pham, T.N.; Nishijo, M.; Pham-The, T.; Tran, N.N.; Vu, H.T.; Tran, A.H.; Tran, T.V.; Nishino, Y.; Nishijo, H. Effect of Perinatal Dioxin Exposure Originating from Agent Orange on Gaze Behavior in 3-Year-Old Children Living in the Most Dioxin-Contaminated Areas in Vietnam. *Toxics* **2022**, *10*, 150. https://doi.org/10.3390/toxics10040150

Academic Editor: Sunmi Kim

Received: 14 February 2022
Accepted: 19 March 2022
Published: 22 March 2022

1. Introduction

During Operation Ranch Hand by the US military in Vietnam from 1961 to 1971, a large quantity of Agent Orange containing 2,3,7,8-tetrachlorodibenzo-p-dioxin (TCDD), which is the most toxic congener of dioxin, was stored and spilled during mixing and loading at several US airbases. This spillage caused considerable dioxin contamination in the environment and to people living at the airbase. This dioxin contamination of Bien Hoa airbase is the largest scale of contamination of an airbase, particularly from TCDD originating from Agent Orange. In 2012, we recruited 210 pairs of mothers and children living in 10 communities close to Bien Hoa airbase (Bien Hoa birth cohort 2012). We reported that mean TEQ-PCDD/Fs levels were 10.5 pg-TEQ/g lipid for primipara and 8.9 pg-TEQ/g lipid for multipara and significantly higher than those in unsprayed areas (3.5 g-TEQ/g lipid for primipara and 3.0 pg-TEQ/g lipid for multipara) [1] and 5.9 pg-TEQ/g of critical maternal level suggested by EFSA Panel on Contaminants in the Food Chain (CONTAM) in 2018 [2]. Moreover, mean TCDD concentrations in maternal breast milk were 2.6 pg/g lipid and approximately 4–5 times higher in primiparous mothers, and 2.2 pg/g lipid and 7–8 times higher in multiparous mothers compared with those in unsprayed areas (0.57 pg/g lipid for primipara and 0.31 pg/g lipid for multipara) [1].

TCDD is the most typical and powerful endocrine rupture chemical (EDC), whose toxicity is induced mainly via activation of the aryl hydrocarbon receptor (AhR) binding to the xenobiotic response element of the target gene [3]. AhR activation by TCDD during the perinatal period was reported to involve in the pathogenesis in developmental disorders associated with alteration of pituitary hormones including growth hormone and gonadotropins and steroid hormones whose synthesis is regulated by pituitary hormones in animals [4–8] and humans [9–13].

Interestingly, a high level of androgen exposure during the fetal period is suggested to increase the risk for autism spectrum disorder (ASD) [14–17], which is one of the neurodevelopmental disorders suspected to be prevalent with increasing prenatal EDC exposure [18].

To investigate the association between child neurodevelopment and dioxin exposure, particularly TCDD, we followed up with children from the Bien Hoa birth cohort 2012 and additionally collected 78 mother–child pairs in Bien Hoa in 2015 (Bien Hoa birth cohort 2015). We then found that children who were perinatally exposed to TCDD at high concentrations (5.5 pg/g lipid of dioxins in breast milk) showed significantly lower expressive language than that in boys at 2 years of age [19]. We also examined neonatal electroencephalograms (EEGs) in children from the Bien Hoa birth cohort 2015 and found that perinatal TCDD exposure altered EEG power values. This alteration led to poor gaze behavior at 2 years of age as indicated by a decreased fixation duration on the face of a child in a video presented on a screen. These results suggest that prenatal TCDD exposure affects fetal brain development and may lead to a poor communication ability among them [20].

Since a reduced fixation duration on faces has frequently been observed in children who have difficulties in social interactions or poor neurodevelopment, such as children with autism [21–23], we examined gaze behavior for viewing static human facial images in children from the Bien Hoa birth cohort 2012. We reported that perinatal exposure to high TCDD exposure may decrease the percentage of the face fixation duration in girls [24]. At that time, however, the associations between gaze behavior and their neurodevelopmental indices, including autistic behavior as indicated by Autism Spectrum Rating Scales (ASRS) scores, were not investigated at that time. We also did not analyze the associations between gaze behavior and head circumference, which are larger in children with autism [25–27] or smaller in girls with autism [28] compared with those in normally developed children or girls.

Therefore, in the present study, we first confirmed the effect of perinatal dioxin exposure, particularly TCDD exposure, on gaze behavior in our previous report [24]. Then, we investigated the associations between gaze behavior and indicators of autistic traits, including ASRS scores and head circumference, in 3-year-old children living in the most dioxin-contaminated areas in Vietnam.

2. Materials and Methods

2.1. Study Areas and Subjects

2.1.1. Study Areas

Bien Hoa airbase is one of the largest former US airbases examined dioxins and heavy dioxin contamination with high TCDD contribution to total dioxins (>80%) was reported in soil and sediment of canals and ponds in the airbases with 61,400 (pg/g dry wet) for the highest concentration of TCDD in soil samples [29]. Shecter et al. (2001) reported higher TCDD in milk samples collected from residents nearby Bien Hoa airbase with 133–1832 (pg/g lipid) in 1970–1973, 2.1–11 (pg-g lipid) in 1985–1988, and 2.0–164 (pg/g lipid) in 1999 compared with the unsprayed area (<2 pg/g lipid) [30].

2.1.2. Study Subjects

In 2012, 224 mother–child pairs living in 10 communities close to Bien Hoa airbase were recruited for a baseline study between September and December (Bien Hoa birth cohort 2012). The mothers delivered their newborns at Dong Nai Prefectural Hospital in Bien Hoa City, Vietnam. The following criteria were used for inclusion in the study: (i) the mothers resided in the target area at least during pregnancy; (ii) the newborns were born full-term, and (iii) there were no complications during birth [1]. A summary of follow-up surveys of the present cohort was shown in Table 1.

Table 1. Summary of Bien Hoa birth cohort 2012.

Year	Age	Performance	N	Examinations
2012	at birth	Recruitment at the hospital	224	Body sizes
	1 month	Maternal breast milk collection	210	Body sizes
2014	2 years	Follow-up	174 *	Body sizes and neurodevelopment
2015	3 years	Follow-up	193	Body sizes and neurodevelopment
			153	Gaze behavior
			142	For present analysis

N: number of subjects. *: Pham TN reported the results of 227 children including these 174 children from the 2012 cohort and 53 children from the 2015 cohort.

A follow-up study at 3 years of age was carried out between November and December 2015. A total of 193 (86.2%) children participated in a survey to examine gaze behavior and their neurodevelopment. However, gaze data from 40 children were missing. Eight children (5 boys and 3 girls) could not attend the gaze test because the electricity was cut-off on the examination day at one of the local survey sites in the community. Eighteen children (5 boys and 13 girls) refused to participate in the gaze test or did not have a successful calibration of the eye-tracking system before the gaze test. Fourteen children (7 boys and 7 girls) showed a lack of attention because the total duration of fixation on the screen was <1.13 s (10th percentile), and they were excluded from the data set. Moreover, 11 children were excluded because of missing data as follows. Seven children did not have dioxin concentrations measured in breast milk because of an insufficient volume of breast milk samples. Two children had missing information regarding the smoking status of family members. Two children were missing ASRS scores because family members who accompanied the children did not know their behavior well. There were no significant differences regarding characteristics or breast milk dioxin concentrations between participants and non-participants or included and excluded subjects. Therefore, the final number of children included for data analysis was 142 (80 boys and 62 girls).

Data on the characteristics of the parents (age, education, family income, primipara/multipara, smoking, and alcohol drinking) and the children (gestational weeks and sex) were collected and are shown in Table 1. The mean maternal age and duration of education with standard deviation (SD) in total participants to the 3-year-old survey in both sexes were 28.5 (4.8) and 11.3 (3.2) years, respectively. The number of primiparous mothers was 55 (38.7%). No mother was a smoker, but 66.9% of mothers stayed with family members who smoked. The proportion of mothers who consumed a product of alcohol during pregnancy was 4.9%, but this was only occasionally (<300 mL of beer each time). The mean gestation was 39.0 weeks and the rate of boys was 56.3%. For these maternal characteristics and family monthly income, there was no significant difference between total cohort pairs and participant pairs (statistic test results were not shown).

Mean values and Z-scores of the children's body size (weight, length/height, body mass index, and head circumference at birth and at 3 years) are shown in Table 1. Body size values were adjusted for age and sex to be standardized Z-scores by following the World Health Organization (WHO) standards (www.who.int/childgrowth/, (accessed on 11 March 2022)). No difference in Z-scores of body size indices at birth was found between total cohort pairs and participant pairs (statistic test results were not shown). Additionally, there was no difference in Z-scores of body sizes at birth and at 3 years of age between boys and girls, except for the head circumference at birth, which showed borderline significance ($p = 0.053$).

Written informed consent was obtained from all of the mothers according to a process that had been reviewed and approved by the Health Department of Dong Nai Province and the Vietnam Military Medical University. The institutional ethics board for epidemiological studies at Kanazawa Medical University approved the study design (No. E-187).

2.2. Dioxin Measurements

When the infants were approximately 1 month old, nurses from a community health station visited the mother's house and collected approximately 20 mL of maternal breast milk. These samples were frozen and transported to Japan inside dry ice by airplane. Approximately 10 mL of breast milk for each sample was used to quantify the levels of 7 congeners of polychlorinated dibenzo-p-dioxins (PCDDs) and 10 congeners of polychlorinated dibenzofurans (PCDFs) in the High Technology Center at Kanazawa Medical University in Uchinada, Japan [31]. After using the EYELA freeze-dryer (FDU-1200; Tokyo-rika Inc., Tokyo, Japan) to freeze and dehydrate breast milk samples, the fat content (g) was extracted using the ASE-200 accelerated solvent extractor (Dionex Co., Sunnyvale, CA, USA). We then added 13C-labeled 2,3,7,8-substituted PCDD/Fs (DF-LCS-A40; Wellington Inc., Guelph, ON, Canada) as an internal standard. A series of procedures, including alkali digestion, hexane extraction, and chromatography on a multilayered silica gel column, were performed to purify samples. A single-layered column of activated carbon was used to separate and collect PCDD/PCDF fractions. Concentrations of 17 PCDD and PCDF congeners were quantified using a gas chromatograph (HP-6980; Hewlett-Packard, Palo Alto, CA, USA) equipped with a high-resolution mass spectrometer (high-resolution-gas chromatography/mass spectrometry, MStation-JMS700; JEOL, Tokyo, Japan) operating in the selected ion-monitoring mode. Toxic equivalent (TEQ)-PCDD/Fs of 7 PCDD and 10 PCDF congeners were calculated as the sum of all values, which were obtained by multiplying each congener concentration by its toxic equivalent factor from WHO 2005-TEF [32]. Congeners that showed a concentration below the detection limit were set to half of the detection limit. The method for analysis has been described in more detail in previous studies [1,33].

TCDD was detected in all samples. Geometrical means and geometrical standard deviations of TCDD and TEQ-PCDD/Fs in maternal breast milk are shown in Table 2. The geometrical mean of TCDD and TEQ-PCDD/F concentrations was 2.5 pg/g lipid and 9.6 pg-TEQ/g lipid, respectively.

2.3. Gaze Behavior Examination

The Tobii X2-60 Compact eye tracker running at 60 Hz (Tobii Technology, Stockholm, Sweden) was used to examine gaze behavior in children aged 3 years. The position of both eyes was recorded by an infrared camera below the screen of a computer. Children were comfortably seated on the lap of their mother or caregiver with a distance from the child's eyes to the screen of approximately 60 cm and with free viewing. At first, 10 static images showing animals and some familiar objects, such as a cake, bicycle, and a pair of scissors, were shown for children as a practice phase. This viewing was performed to attract attention from children toward the screen. Ten static images of the human face were then shown to children as gaze stimuli, with 3 s for each picture. Before recording, children had to be successfully calibrated by the eye-tracking system with five red points, which appeared at the center and four corners of the screen.

Table 2. Characteristics of mothers and children and perinatal dioxin concentrations.

		All Cohort	Total (N = 142)	Boys (N = 80)	Girls (N = 62)
Characteristics	**Units**	**Mean (SD), N (%)**	**Mean (SD), N (%)**	**Mean (SD), N (%)**	**Mean (SD), N (%)**
Mothers					
Age	years	28.5 (4.6)	28.5 (4.8)	28.9 (4.9)	27.9 (4.6)
Education	years	11.3 (3.1)	11.3 (3.2)	11.4 (3.3)	11.2 (3.3)
Income (per a month)	million VND	9.8 (11.3)	10.2 (13.5)	9.6 (5.2)	11.0 (19.6)
Parity (% primipara)	N (%)	82 (37.6)	55 (38.7)	29 (36.3)	26 (41.9)
Alcohol drinking	N (%)	10 (4.7)	7 (4.9)	6 (7.5)	1 (1.6)
Family smoking	N (%)	140 (65.4)	95 (66.9)	50 (62.5)	45 (72.6)
Children					
Gender (rate of boys)	N (%)	116 (53.2)	80 (56.3)		
Gestational period	weeks	39.0 (1.2)	39.0 (1.3)	39.0 (1.3)	39.0 (1.2)
At birth					
Weight	g	3297 (411)	3272 (416)	3353 (425)	3168 (383)
	Z-score	0.01 (0.82)	−0.05 (0.82)	0.014 (0.83)	−0.13 (0.80)
Length	cm	49.9 (2.2)	49.8 (2.1)	50.1 (2.0)	49.5 (2.2)
	Z-score	0.18 (1.19)	−0.04 (2.48)	0.12 (1.04)	−0.25 (3.58)
Head circumference	cm	34.2 (3.4)	34.0 (3.8)	34.5 (3.2)	33.4 (1.8)
	Z-score	0.12 (2.10)	0.03 (2.18)	0.34 (2.54)	−0.38 (1.51)
BMI		13.3 (1.4)	13.2 (1.4)	13.3(1.4)	13.0 (1.3)
	Z-score	−0.09 (1.03)	−0.15 (1.04)	−0.05 (1.08)	−0.27 (0.99)
At 3 years old					
Age	month		37.5 (0.7)	37.5 (0.8)	37.6 (0.6)
Weight #	kg		15.2 (2.7)	15.5 (2.8)	14.7 (2.6)
	Z-score		0.41 (1.35)	0.52 (1.47)	0.26 (1.17)
Height #	cm		95.1 (3.5)	95.3 (3.5)	94.9 (3.5)
	Z-score		−0.32 (0.92)	−0.38 (0.94)	−0.23 (0.89)
Head circumference #	cm		48.5 (1.5)	49.0 (1.4)	47.7 (1.3)
	Z-score		−0.48 (0.95)	−0.38 (0.98)	−0.60 (0.91)
BMI #			16.7 (2.2)	16.9 (2.2)	16.2 (2.1)
	Z-score		0.88 (1.65)	1.07 (1.75)	0.63 (1.50)
Breast milk					
TCDD	GM (GSD)	2.3 (2.3)	2.4 (2.3)	2.4 (2.2)	2.4 (2.5)
TEQ-PCDD/Fs	GM (GSD)	9.1 (1.7)	9.4 (1.7)	9.5 (1.7)	9.3 (1.7)

N: number of subjects; SD: standard deviation; VND: Vietnam Dong per month; alcohol drinking: drinking habits during pregnancy; GM: geometrical mean; GSD: geometrical standard deviation; BMI: body mass index; Z-score: adjusted by age and sex with WHO reference; TCDD: 2,3,7,8-tetrachlorodibenzo-p-dioxin, TEQ-PCDD/Fs: toxic equivalency of polychlorinated dibenzo-p-dioxins and polychlorinated dibenzo furans; #: missing for 1 boy and 1 girl.

We used Tobii Studio version 3.3.1 (Tobii Technology, Stockholm, Sweden) to extract children's data and analyze their fixation duration. Before extracting data, the default I-VT filter was applied with the velocity threshold and minimum fixation duration set at 30 degrees/s and 100 ms, respectively, to determine fixation. Figure 1 shows the fixation density of the child when they viewed a picture. Whole picture and face areas were manually set to define areas of interest in each picture. The total fixation duration on the whole picture and face areas in each picture (by second) were extracted by Tobii software. The total fixation duration on pictures and face areas was calculated by adding up these values for 10 pictures. The percentage of the total fixation duration on faces was defined as the ratio of the total fixation duration on faces in all pictures divided by the total fixation duration (s) in all pictures, and then the ratio was multiplied by 100 to obtain the percentage of face fixation. This method was described in more detail in our previous studies [20,24,34].

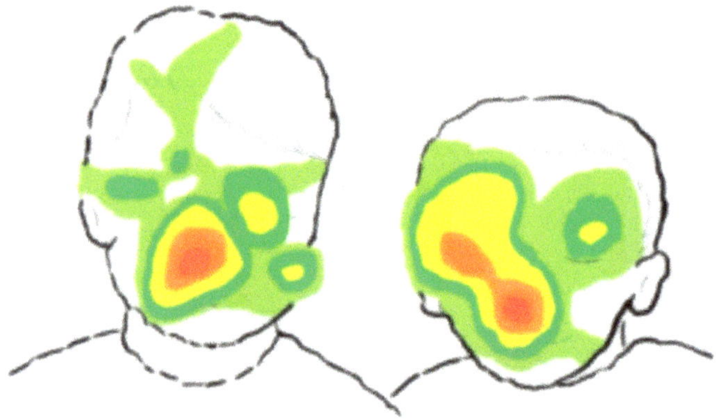

Figure 1. Density of the fixation duration when children watched a picture of two children.

2.4. Neurodevelopmental Assessment

General neurodevelopment was examined using the Bayley III (NCS Pearson, Inc., Bloomington, MN, USA) across five domains of cognition, expressive and receptive language, and fine and gross motor skills. Two examiners performed this test in our previous study [19], who were trained well and blinded to dioxin exposure levels of the children who were administered the tests. At this time, cognition and language ability were assessed by one examiner, and another examiner assessed motor development in all children.

Full-length parent rating forms (2–5 years) of the ASRS (MSH, North Tonawanda, NY, USA) with 70 questions that assess child behavior associated with autism spectrum disorder (ASD) were used to interview mothers or caregivers of children in this study. Before the survey, a trial examination was conducted on a group of 15 Vietnamese children to ensure the feasibility and appropriateness of the ASRS for the Vietnamese population [35]. The total score, the Diagnostic and Statistical Manual of Mental Disorders, 4th edition, Text Revision (DSM-IV-TR; DSM) score, the social communication score, and the Unusual Behavior score were calculated after the interview by following the ASRS guidebook (ASRS, Technical Manual). The total score and DSM scores were >60, which is suspected to have autistic traits. One medical doctor was trained well by a specialist and interviewed all subjects in this study. The detailed method was described in a previous study [35].

2.5. Statistical Analysis

The IBM SPSS (ver. 21) software package for Windows (IBM Corp., Armonk, NY, USA) was used for statistical analysis. Concentrations of 17 PCDD/F congeners and TEQ-PCDD/Fs in breast milk were logarithmically transformed (base 10) to improve normality. A general linear model was used to compare the face fixation duration (%) between the high and low exposure groups after adjusting covariates. The cut-off value for high and low TCDD concentrations was 3.5 pg/g lipid. This value was based on the geometrical mean and the geometrical SD of dioxin concentrations in breast milk samples from unsprayed areas reported in our previous study using the following equation: geometrical mean × geometrical SD^3 [36]. For the cut-off value of TEQ-PCDD/Fs, the 75th percentile value of TEQ-PCDD/Fs in the subjects was used because of comparatively lower concentrations than those in another dioxin hotspot around Da Nang airbase [33,36]. The following factors were included as covariates: maternal age and education (years), parity (primipara/multipara), alcohol consumption during pregnancy (yes/no), family income, family member smoking (yes/no), and gestational weeks. To analyze the dose–effect relationship between TCDD concentrations and the face fixation duration (%), subjects were divided into three groups (low, middle, and high) with cut-off values of 3.5 and 6.3 pg/g lipid. At this time, the 90th percentile value of TCDD in these subjects was used for the higher cut-off value. Differences among the groups were analyzed using the general linear model.

For analyses of associations between the face fixation duration (%) and neurodevelopmental indices or body size, including the head circumference, we used the linear regression model after adjusting for the same covariates as those used in general linear model analysis. For all analyses, significance was considered $p < 0.05$.

3. Results

3.1. Effects of Perinatal TCDD and TEQ-PCDD/F Exposure on Gaze Behavior

We compared the adjusted mean face fixation duration (%) between the high and low exposure groups according to TCDD or TEQ-PCDD/F concentrations in boys and girls (Table 3). No significant difference in the face fixation duration (%) was found between the high and low TCDD exposure groups in boys. However, in girls, the adjusted mean face fixation duration (%) was significantly lower ($p < 0.05$) in the high TCDD group compared with that in the low TCDD group. However, no significant difference in the face fixation duration (%) was found between the high and low TEQ-PCDD/F exposure groups in both sexes (Table 3).

Table 3. Comparison of the adjusted mean face fixation duration (%) between the high and low dioxin exposure groups.

	Low Exposure Group				High Exposure Group				
	N	Mean	adj. Mean	95% CI (Lower, Upper)	N	Mean	adj. Mean	95% CI (Lower, Upper)	*p*-Value
Boys									
TCDD	62	48.0	46.6	(40.8, 52.4)	18	46.6	51.3	(39.8, 62.8)	0.483
TEQ-PCDDs/Fs	60	45.7	45.6	(39.7, 51.5)	20	53.7	53.8	(43.2, 64.5)	0.194
Girls									
TCDD	45	43.1	43.4	(35.8, 50.9)	17	26.3	25.5	(12.7, 38.3)	0.025
TEQ-PCDDs/Fs	46	40.1	40.2	(32.5, 47.8)	16	33.8	33.6	(20.1, 47.0)	0.410

TCDD: 2,3,7,8-tetrachlorodibenzo-p-dioxin; TEQ: toxic equivalent; PCDD/Fs: polychlorinated dibenzo-p-dioxins and polychlorinated dibenzo furans; cut-off values: 3.5 (pg/g lipid) for TCDD, 12.5 (pg-TEQ/g lipid) for TEQ-PCDD/Fs; N: number of subjects; adj.Mean: adjusted mean, 95% CI: 95% confidence interval; covariates: age, education, parity, and drinking during pregnancy of mothers, family income, family members' smoking habit, and gestational weeks, birth weight, and age (months) at the examination of children.

To clarify the dose–effect relationship between perinatal TCDD exposure and gaze performance, the mean face fixation duration (%) was compared among the three TCDD exposure groups of low (<3.5), middle (3.5–6.3), and high (≥6.3) (pg/g lipid), after adjusting for covariates (Figure 2). In girls, the adjusted mean face fixation duration (%) significantly decreased ($p < 0.05$) as exposure levels of TCDD increased. However, in boys, no dose–response relationship was found between TCDD concentrations and the face fixation duration (%).

At this time, we did not analyze the dose–effect relationship with TEQ-PCDD/Fs exposure because the rates of high TEQ-PCDD/Fs (≥12.5 pg-TEQ/g lipid) were 10.3, 52.6, and 93.7 (%) for low, middle, and high TCDD groups, suggesting TCDD is a good exposure marker to show a wide range of dose in the present study.

3.2. Gaze Behavior and Neurodevelopment as Indicated by Bayley III and ASRS Scores

To investigate the relationships between gaze behavior and neurodevelopment, including autistic traits, the associations between the face fixation duration (%) and Bayley III and ASRS scores were analyzed using the regression linear model. An adjustment was made for the same covariates that were used in the general linear model analysis (Table 4). There was no significant association between the face fixation duration (%) and Bayley III scores in either sex. However, in girls, the face fixation duration (%) increased with an increase in composite and receptive language scores (β = 0.293 and β = 0.245, respectively; $p = 0.058$ and $p = 0.064$, respectively).

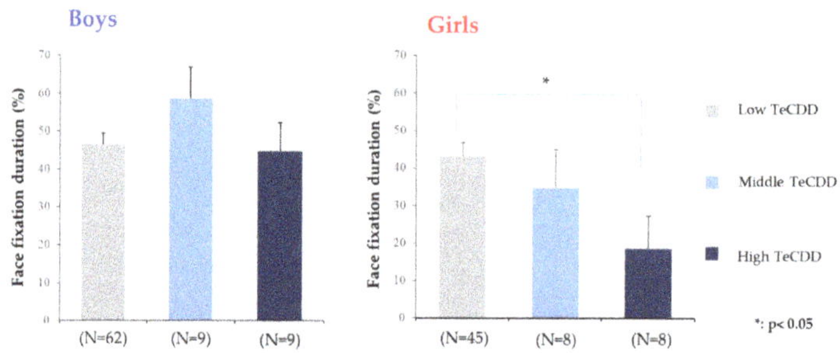

Figure 2. The dose–effect relationship between perinatal TCDD exposure and gaze performance: comparison of the adjusted face fixation duration (%) among the low (<3.5), middle (3.5–6.3), and high (≥6.3 pg/g lipid) TCDD groups in boys and girls. Note: TCDD: 2,3,7,8-tetrachlorodibenzo-p-dioxin; N: number of subjects.

Table 4. Associations between the face fixation duration (%) and neurodevelopmental scale scores after adjusting for covariates.

	Boys				Girls			
	N	β	95% CI (Lower, Upper)	p-Value	N	β	95% CI (Lower, Upper)	p-Value
Bayley III								
Congnition	80	−0.081	(−0.381, 0.220)	0.594	59	0.137	(−0.140, 0.360)	0.380
Language (composite)	77	0.010	(−0.208, 0.226)	0.934	56	0.293	(−0.011, 0.653)	0.058
Receptive language	79	0.021	(−0.203, 0.242)	0.864	57	0.245	(−0.019, 0.616)	0.064
Expressive language	77	0.043	(−0.171, 0.247)	0.717	56	0.236	(−0.083, 0.599)	0.135
Motor (composite)	78	0.166	(−0.063, 0.350)	0.170	60	0.091	(−0.234, 0.448)	0.533
Fine Motor	80	0.113	(−0.110, 0.309)	0.348	61	−0.079	(−0.439, 0.258)	0.604
Gross Motor	78	0.159	(−0.069, 0.350)	0.184	60	0.195	(−0.111, 0.556)	0.186
ASRS								
Social communication	80	−0.132	(−0.327, 0.092)	0.267	62	−0.290	(−0.637, −0.019)	0.038
Unusual behavior	80	0.022	(−0.193, 0.233)	0.854	62	0.027	(−0.287, 0.347)	0.852
Total score	80	−0.066	(−0.263, 0.147)	0.574	62	−0.193	(−0.547, 0.101)	0.173
DSM-score	80	−0.031	(−0.234, 0.179)	0.792	62	−0.224	(−0.574, 0.060)	0.110

N: number of subjects; β: standardized beta; 95% CI: 95% confidence interval; Bayley III: the Bayley Scales of Infant and Toddler Development, Ver. 3; ASRS: autism spectrum rating scale; DSM: the DSM-IV-TR score; covariates: age, education, parity, and drinking during pregnancy of mothers, family income, family members' smoking habit, and gestational weeks, birth weight, and age (months) at the examination of children.

No significant association of ASRS scores was observed in boys. In girls, the face fixation duration (%) was inversely and significantly associated ($p < 0.05$) with social communication scores, where a lower face fixation duration (%) was associated with higher social communication scores (poor social communication skills).

We also compared the adjusted mean Bayley III and ASRS scores between the high and low TCDD groups using the general linear model. However, there were no significant differences in any scale scores of the ASRS or Bayley III between the two TCDD exposure groups in girls (data not shown).

3.3. *Gaze Behavior and Head Circumference at Birth and 3 Years of Age*

The associations between the face fixation duration (%) and all body size indices (Z-score), which comprised weight, length/height, head circumference, and body mass index, at birth and at 3 years of age were analyzed after adjusting for covariates in each sex (Table 5). In boys, there were no significant associations between the face fixation duration

(%) and any body size indices at birth and at 3 years of age. However, the face fixation duration (%) was significantly and inversely associated ($p < 0.05$) with height at 3 years of age in girls. The face fixation duration (%) appeared to increase as the head circumference increased (β = 0.230), but this association was not significant ($p = 0.125$).

Table 5. Associations between the face fixation duration (%) and standardized body size indices (Z-scores) after adjusting for covariates.

	Boys				Girls			
	N	β	95% CI (Lower, Upper)	*p*-Value	N	β	95% CI (Lower, Upper)	*p*-Value
At birth								
Weight	80	0.236	(−0.030, 0.473)	0.085	62	−0.115	(−0.415, 0.173)	0.412
Lenght	80	0.210	(−0.118, 1.074)	0.115	62	−0.185	(−0.334, 0.070)	0.195
BMI	80	0.097	(−0.139, 0.317)	0.438	62	0.113	(−0.194, 0.439)	0.440
Head circuference	80	0.075	(−0.129, 0.252)	0.522	61	0.051	(−0.366, 0.517)	0.733
At 3 years of age								
Weight	79	0.026	(−0.181, 0.227)	0.824	61	−0.110	(−0.495, 0.232)	0.472
Height	79	0.024	(−0.217, 0.262)	0.852	61	−0.292	(−0.612, −0.013)	0.041
BMI	79	0.029	(−0.179, 0.232)	0.799	61	0.037	(−0.316, 0.400)	0.814
Head circuference	79	0.027	(−0.192, 0.242)	0.817	61	0.230	(−0.071, 0.568)	0.125
Head circumference *	78	0.038	(−0.197, 0.267)	0.765	61	0.281	(0.004, 0.605)	0.047

N: number of subjects; β: standardized beta; 95% CI: 95% confidence interval; BMI: body mass index; covariates at birth: age, education, parity, and drinking during pregnancy of mothers, family income, family me mbers' smoking habit, and gestational weeks at birth; covariates at 3 years of age: age, education, parity, and drinking during pregnancy of mothers, family income, family members' smoking habit, and gestational weeks, birth weight, and age (months) at the examination of children; *: height at 3 years of age was added to covariates above.

Because the head circumference was correlated with height, a regression analysis between the face fixation duration (%) and head circumference at 3 years of age was performed again after adjusting for covariates, including height at 3 years of age, in both sexes (Table 5). The association between the face fixation duration (%) and head circumference at 3 years of age was significantly increased after adjusting for height (β = 0.281, $p < 0.05$), which suggested that children with a lower gaze fixation (%) on the face had a smaller head size.

4. Discussion

4.1. Effect of Perinatal Dioxin Exposure on Gaze Behavior in 3-Year-Old Children

In girls aged 3 years from the Bien Hoa cohort 2012, the gaze fixation duration on faces of static pictures decreased as perinatal TCDD exposure increased in a dose–effect manner. This finding suggested that atypical gaze behavior was associated with TCDD exposure in girls. However, no association was observed between dioxin congeners other than TCDD and gaze behavior in either sex, which indicated that TCDD is the only congener that affects the gaze behavior of children.

Gaze behavior has been frequently examined in children with autism or poor general neurodevelopment, and they have atypical gaze behavior with fewer fixations on the whole face compared with children without these conditions [21–23]. However, previous studies on the effects of organochlorine compounds on gaze behavior are limited, except for one report on polychlorinated biphenyl congeners and gaze behavior in Japanese infants [37]. This Japanese study reported that infants who were prenatally exposed to high concentrations of polychlorinated biphenyl #118 preferred to fix their gaze on inverted point-light displays of a walking human figure rather than reduce attention to upright biological motion.

In the Bien Hoa birth cohort 2015, we found that perinatal TCDD altered neonatal EEG power values, which were associated with a reduced face fixation duration (%) in children aged 2 years when viewing dynamic social stimuli (a movie of a young girl playing in front of her mother taking a video of her) [20]. However, different gaze stimuli in this previous study were used compared with those used in the present study. Moreover, we previously reported that perinatal TCDD exposure may decrease face fixation duration in 3-year-old girls from the Bien Hoa birth cohort 2012 based on the same dataset as that used in the present study [24]. In the present study, we found a clearer dose–response relationship between TCDD exposure and the face fixation duration in girls after excluding more children with inattention compared with that in the previous analysis.

In the present analysis, we analyzed the associations between gaze behavior and concentrations of PCDD congeners other than TCDD, but no significant difference was observed between the high and low exposure groups (data not shown). We also found that autistic traits increased with an increase in TCDD concentrations, but not TEQ-PCDD/Fs, in 3-year-old children in our previous study in children from the Da Nang birth cohort [35]. However, Nowack et al. (2015) reported an effect of prenatal exposure to TEQ-PCDD/Fs on autistic traits in childhood using the Social Responsiveness Scale in children from the Duisburg birth cohort in Germany [38]. Therefore, we need to follow up with these children to determine whether TCDD is a specific congener that increases autistic behavior, including gaze behavior, on faces in the future.

4.2. Gaze Behavior and Child Neurodevelopment including Autistic Traits

In the present study, in girls, the face fixation duration (%) significantly decreased as social communication scores of the ASRS increased. This finding suggested poor social communication skills, which are often observed in children with autism, although no direct association was found between TCDD exposure and ASRS scores. The face fixation duration (%) also tended to decrease as Bayley III language scores decreased in girls, which suggested that a decreased gaze fixation on faces indicated a poor communication ability. These results suggest that a decreased face fixation duration may be associated with increased autistic traits in children from the Bien Hoa birth cohort 2012.

Gaze behavior toward faces, particularly the eyes and mouth regions, has been investigated using eye-tracking of static pictures or video clips in individuals with autism, because of social communication deficit which is one of their characteristic symptoms [39]. Previous studies showed decreased gaze behavior in the faces indicated by shorter fixation duration on faces in children [21,23] and adults [22] with ASD compared with normally developed children and adults. Particularly in the eye regions, decreased gaze behavior was found in adults with ASD [22] but decreased gaze behavior was observed in the mouth regions in children with ASD [23].

We also analyzed gaze behavior in the eyes and mouth in the present study but found no significant difference in gaze behavior in the eyes or mouth regions between the high and low TCDD groups (data not shown). Nakano et al. (2010) used video clips with sound involving human characters who were talking to the audience or each other [23]. This type of dynamic stimulus may be more attractive for children and induce different gaze behavior from that for static pictures because of viewing a talking mouth and face and hearing voices. In the future, we will analyze the fixation duration on the face, eyes, and mouth regions after separating talking scenes from silent scenes, and clarify the atypical gaze pattern associated with dioxin exposure in Vietnamese children.

The risk of ASD in childhood is suspected to be increased by high androgen exposure during the fetal period [14–17]. It was also reported that aromatase expression and estrogen or estrogen receptor expression were associated with the development of ASD [40,41]. On the other hand, TCDD is the most powerful EDC which may alter the uterine endocrinal environment, particularly pituitary hormones including growth hormone and gonadotropins, and influence fetal development [4–8]. Taken together, perinatal TCDD exposure may influence the fetal neuroendocrine environment which contributes to pathology for ASD, leading to increased autistic traits in children exposed to TCDD in early life.

4.3. Gaze Behavior and Head Circumference

In the present study, the face fixation duration (%) decreased with an increase in height at 3 years of age in girls. This finding suggested that girls who had a shorter gaze fixation on faces were taller. In contrast, the face fixation duration (%) significantly decreased with a decrease in the head circumference at 3 years of age after adjusting for confounding factors including height in girls. This finding suggested that girls with a shorter fixed gaze on faces had a smaller head size relative to their height.

An increased rate of macrocephaly indicating a large brain volume has been suggested in children with autism in previous studies [25–27]. Lainhart et al. (2006) reported that the mean standardized head size was significantly increased relative to height in individuals with autism. They also investigated the distribution of height, which is correlated with the head circumference in autism, and showed no difference in height between individuals with and those without autism [27].

A recent meta-analysis of 12 studies included autistic and normally developed children [28]. This analysis showed that in girls with autism, the mean head circumference was smaller than that in normally developed girls at 12–17 months of age. Additionally, an extreme head circumference above and below 1.5 standard deviations was reported when autistic girls reached 36–59 months of age. However, there was no significant difference in the mean head circumference and extreme head circumference at the same age ranges in boys with autism. These results suggest that the head size may be affected by age and sex in autism and that the rate of a smaller head circumference may be more frequent at 3 years of age in girls. Therefore, 3-year-old girls who were perinatally exposed to high TCDD concentrations in the present study may have had a smaller head associated with increased autistic traits. This possibility suggests that changes in brain development are associated with TCDD exposure during the perinatal period.

In the present study, a shorter gaze fixation on faces was more likely in taller girls. Almost no evidence was found between gaze behavior, and autistic traits and height in previous literature, except for a study on children with high-functioning autism and Asperger disorder [42]. This previous study reported a high growth rate of length/height during the first 3 years of life in these children. In the present study, no effect of dioxin exposure on height was found in children of either sex (data not shown). Therefore, we need to follow up this birth cohort in the future to clarify the relationship between gaze behavior/autistic traits and body size parameters, including head circumference and height, in each sex.

4.4. Sex Differences in Dioxin Effects on Neurodevelopment and Behavior in Children

In a neurodevelopmental study at 2 years of age in children from the Bien Hoa birth cohorts 2012 and 2015 including the present subjects, a significantly lower expressive language score was associated with perinatal TCDD exposure only in boys [2]. Similarly, in follow-up studies in children from the Da Nang cohort, a significant decrease in motor and expressive language scores associated with perinatal TCDD and TEQ-PCDD/F exposure in the first 3 years of life [43] and decreased scores of cognitive ability and motor coordination skills associated with increased TEQ-PCDD/Fs at 5 years of age [44] were found only in boys. Poor learning ability was found only in 8-year-old boys exposed to high PCDD congener concentrations, including TCDD, during the perinatal period [45].

However, another study of children in Da Nang showed that increased ADHD symptoms, particularly impulsivity and hyperactivity, were associated with TCDD exposure only in girls [46]. In a previous study to examine feminine visual preference using eye-tracking of pictures of boy- or girl-oriented objects, we reported that visual interest in girl-oriented pictures was associated with high TCDD exposure in girls, and associated with PCDD congeners other than TCDD in boys [34]. Moreover, Vu et al. (2021) reported that mirror neuron activity indicating a reduction in EEG power by observation of hand movements, which plays an important role in social–emotional behavior, was altered with high perinatal TCDD exposure in girls at 9 years of age [47]. These results suggest that perinatal TCDD exposure might affect specific domains of neurodevelopment, such as social–emotional behavior, in older girls (early school age).

However, in most previous studies that investigated the association between dioxin exposure and neurodevelopment in children at similar ages in countries other than Vietnam, effects of dioxins were reported only in boys. In the Seveso Second Generation Health Study in Italy, Ames et al. (2019) reported that increased non-perseverative errors of the Wisconsin card sorting test were associated with perinatal TCDD exposure only in boys at 7–17 years of age [48]. In the Rotterdam cohort in The Netherlands, Vreugdenhil et al. (2002) reported that perinatal TEQ-PCDD/Fs exposure increased feminine behavior, which was more pronounced in boys than in girls at 8 years old [49]. In the Duisburg cohort in Germany, perinatal TEQ-PCDD/Fs exposure affected sex-typed behavior both in boys and girls aged 6–8 years [50].

These results indicating sex-patterned alteration of child behavior due to perinatal dioxin exposure might be related to characteristics of EDCs which alternately show estrogenic and androgenic effects in a sex-specific manner.

During the perinatal period, infants are suggested to be highly hormone-sensitive and influenced by TCDD exposure even at lower levels than the levels to induce systemic toxicity in adults based on animal studies [51,52] (bimodal effects of EDCs). In animal models, perinatal TCDD exposure at a low level reduced fetal luteinizing hormone (LH) levels by downregulation in the pituitary gene, leading to impaired sexual behavior in later life [53]. Since AhR has an important role in the homeostatic regulation of LH production in the pituitary gland [54], TCDD binding AhR is suggested to disrupt LH production and induce sexual immaturity in fetuses [55].

Taken together, we need to follow up our Vietnamese birth cohorts exposed to TCDD until adolescence when their hormonal environment will be drastically changed, and to investigate their pituitary hormonal levels and puberty arrival in the future.

4.5. Limitations

In this study, we investigated the effects of perinatal TCDD exposure on gaze behavior using static pictures as gaze stimuli. However, static pictures are likely to have a less natural form and motion than dynamic stimuli in a video clip and may induce different gaze behavior from that with watching faces in real life [56]. Therefore, further studies on gaze behavior in children using dynamic facial stimuli shown in a video clip are required.

Another limitation to the present study is its relatively small sample size because of the high rate (18.4%; 32 children/172 participants) of an unsuccessful gaze behavior examination. In particular, calibration with viewing five red points before recording was difficult for some children. A dynamic cartoon figure accompanied by sound for calibration instead of red points could be used to easily attract the children's attention. Additionally, changing stimuli from static pictures to a movie of playing and talking children may improve the point of eye-tracking to make young participants pay more attention to face stimuli and achieve sufficient gaze behavior for analysis.

Eye tracking, which we used in the present study, is an objective tool for assessing gaze behavior to clarify children's preferences and interests, which are not able to be obtained by a questionnaire survey. Further study using the eye-tracking method with a greater number of subjects, including unexposed control children, is necessary to clarify the association between dioxin exposure and atypical behavior in children from birth cohorts in Vietnam.

5. Conclusions

This study showed that 3-year-old girls living in a hot spot of dioxin contamination showed atypical gaze behavior when watching children's faces in pictures that were proportional to perinatal TCDD exposure. These girls with atypical gaze behavior showed lower social communication scores and had a smaller head size, which can be found in girls with autism. We will follow up this birth cohort and investigate them again with an improved eye-tracking method to clarify the effects of perinatal dioxin exposure, particularly TCDD, on their behavior and hormonal levels in late childhood and adolescence in the future.

Author Contributions: Conceptualization, M.N., Y.N. and H.N.; methodology, T.N.P., T.P.-T. and M.N.; investigation, T.N.P., M.N., T.P.-T., N.N.T., A.H.T., T.V.T. and H.T.V.; data curation, T.N.P.; writing—original draft preparation, T.N.P. and M.N.; writing—review and editing, Y.N. and H.N.; funding acquisition, M.N. All authors have read and agreed to the published version of the manuscript.

Funding: This work was partly supported by the Ministry of Education, Science, Sports and Culture with a Grant-in-Aid for Scientific Research (B) (25305024 and 17H04665). The funders had no role in the study design, data collection and analysis, decision to publish, or preparation of the manuscript.

Institutional Review Board Statement: The study was conducted according to the guidelines of the Declaration of Helsinki and approved by the Institutional Review Boards of the Health Department of Dong Nai City and the Kanazawa Medical University (No. E-187, 25 December 2013, updated in 17 January 2017).

Informed Consent Statement: Informed consent was obtained from all subjects involved in the study.

Data Availability Statement: The data presented in this study are available on request to the corresponding author. The data are not publicly available due to the personal information (gaze data).

Acknowledgments: We would like to thank all mothers and children who participated in this study. We also thank Phan Huy Anh Vu and Huynh Tu Anh Dong Nai in the Health Department, Nguyen Xuan Hung, and Nguyen Thi Phuong Thao in Bien Hoa Health Center, and medical staff in the 10 communities close to Bien Hoa airbase for their collaboration.

Conflicts of Interest: The authors declare no conflict of interest.

References

1. Nghi, T.N.; Nishijo, M.; Manh, H.D.; Tai, P.T.; Van Luong, H.; Anh, T.H.; Thao, P.N.; Trung, N.V.; Waseda, T.; Nakagawa, H.; et al. Dioxins and Nonortho PCBs in Breast Milk of Vietnamese Mothers Living in the Largest Hot Spot of Dioxin Contamination. *Environ. Sci. Technol.* **2015**, *49*, 5732–5742. [CrossRef] [PubMed]
2. EFSA Panel on Contaminants in the Food Chain (CONTAM); Knutsen, H.K.; Barregård, J.A.L.; Bignami, M.; Brüschweiler, B.; Ceccatelli, S.; Cottrill, B.; Dinovi, M.; Edler, L.; Grasl-Kraupp, B.; et al. Risk for animal and human health related to the presence of dioxins and dioxin-like PCBs in feed and food. *EFSA J.* **2018**, *16*, e05333.
3. Fernandez-Salguero, P.M.; Hilbert, D.M.; Rudikoff, S.; Ward, J.M.; Gonzalez, F.J. Arylhydrocarbon receptor-deficient mice are resistant to 2,3,7,8-tetrachlorodibenzo-pdioxin-induced toxicity. *Toxicol. Appl. Pharmacol.* **1996**, *140*, 173–179. [CrossRef] [PubMed]
4. Mably, T.A.; Moore, R.W.; Goy, R.W.; Peterson, R.E. In utero and lactational exposure to 2,3,7,8-tetrachlorodibenzo-p-dioxin. *Toxicol. Appl. Pharmacol.* **1992**, *114*, 108–117. [CrossRef]
5. Gray, L.E., Jr.; Kelce, W.R.; Monosson, E.; Ostby, J.S.; Birnbaum, L.S. Exposure to TCDD during development permanently alters reproductive function in male Long Evans rats and hamsters: Reduced ejaculated and epididymal sperm numbers and sex accessory gland weights in offspring with normal androgenic status. *Toxicol. Appl. Pharmacol.* **1995**, *131*, 108–118. [CrossRef]
6. Gray, L.E., Jr.; Wolf, C.; Mann, P.; Ostby, J.S. In utero exposure to low doses of 2,3,7,8-tetrachlorodibenzo-p-dioxin alters reproductive development of female Long Evans hooded rat offspring. *Toxicol. Appl. Pharmacol.* **1997**, *146*, 237–244. [CrossRef]
7. Takeda, T.; Fujii, M.; Hattori, Y.; Yamamoto, M.; Shimazoe, T.; Ishii, Y.; Himeno, M.; Yamada, H. Maternal exposure to dioxin imprints sexual immaturity of the pups through fixing the status of the reduced expression of hypothalamic gonadotropinreleasing hormone. *Mol. Pharmacol.* **2014**, *85*, 74–82. [CrossRef]
8. Takeda, T.; Taura, J.; Hattori, Y.; Ishii, Y.; Yamada, H. Dioxin-induced retardation of development through a reduction in the expression of pituitary hormones and possible involvement of an aryl hydrocarbon receptor in this defect: A comparative study using two strains of mice with different sensitivities to dioxin. *Toxicol. Appl. Pharmacol.* **2014**, *278*, 220–229.
9. Guo, Y.L.; Hsu, P.C.; Hsu, C.C.; Lambert, G.H. Semen quality after prenatal exposure to polychlorinated biphenyls and dibenzofurans. *Lancet* **2000**, *356*, 1240–1241. [CrossRef]
10. Tsukimori, K.; Uchi, H.; Tokunaga, S.; Yasukawa, F.; Chiba, T.; Kajiwara, J.; Hirata, T.; Furue, M. Blood levels of PCDDs, PCDFs, and coplanar PCBs in Yusho mothers and their descendants: Association with fetal Yusho disease. *Chemosphere* **2013**, *90*, 1581–1588. [CrossRef]
11. Mocarelli, P.; Gerthoux, P.M.; Needham, L.L.; Patterson, D.G., Jr.; Limonta, G.; Falbo, R.; Signorini, S.; Bertona, M.; Crespi, C.; Sarto, C.; et al. Perinatal exposure to low doses of dioxin can permanently impair human semen quality. *Environ. Health Perspect.* **2011**, *119*, 713–718. [CrossRef] [PubMed]

12. Manh, H.D.; Kido, T.; Okamoto, R.; Xianliang, S.; Viet, N.H.; Nakano, M.; Tai, P.T.; Maruzeni, S.; Nishijo, M.; Nakagawa, H.; et al. The relationship between dioxins and salivary steroid hormones in Vietnamese primiparae. *Environ. Health Prev. Med.* **2013**, *18*, 221–229. [CrossRef]
13. Warner, M.; Rauch, S.; Ames, J.; Mocarelli, P.; Brambilla, P.; Signorini, S.; Eskenazi, B. Prenatal dioxin exposure and thyroid hormone levels in the Seveso second generation study. *Environ. Res.* **2020**, *183*, 109280. [CrossRef] [PubMed]
14. Knickmeyer, R.; Baron-Cohen, S.; Raggatt, P.; Taylor, K. Fetal testosterone, social relationships, and restricted interests in children. *J. Child Psychol. Psychiatry* **2005**, *46*, 198–210. [CrossRef] [PubMed]
15. Auyeung, B.; Knickmeyer, R.; Ashwin, E.; Taylor, K.; Hackett, G.; Baron-Cohen, S. Effects of fetal testosterone on visuospatial ability. *Arch. Sex. Behav.* **2012**, *41*, 571–581. [CrossRef] [PubMed]
16. Auyeung, B.; Baron-Cohen, S.; Ashwin, E.; Knickmeyer, R.; Taylor, K.; Hackett, G. Fetal testosterone and autistic traits. *Br. J. Psychol.* **2009**, *100 Pt 1*, 1–22. [CrossRef]
17. Baron-Cohen, S.; Auyeung, B.; Nørgaard-Pedersen, B.; Hougaard, D.M.; Abdallah, M.W.; Melgaard, L.; Cohen, A.S.; Chakrabarti, B.; Ruta, L.; Lombardo, M.V. Elevated fetal steroidogenic activity in autism. *Mol. Psychiatry.* **2015**, *20*, 369–376. [CrossRef]
18. Silver, M.K.; Meeker, J.D. Chapter 14—Endocrine disruption of developmental pathways and children's health. In *Endocrine Disruption and Human Health*, 2nd ed.; Darbre, P.D., Ed.; Elsevier: Berlin/Heidelberg, Germany, 2021; pp. 291–320.
19. Pham, N.T.; Nishijo, M.; Pham, T.T.; Tran, N.N.; Le, V.Q.; Tran, H.A.; Phan, H.; Nishino, Y.; Nishijo, H. Perinatal dioxin exposure and neurodevelopment of 2-year-old Vietnamese children in the most contaminated area from Agent Orange in Vietnam. *Sci Total Environ.* **2019**, *678*, 217–226. [CrossRef]
20. Pham, N.T.; Nishijo, M.; Nghiem, T.; Pham, T.T.; Tran, N.N.; Le, V.Q.; Vu, T.H.; Tran, H.A.; Phan, H.; Do, Q.; et al. Effects of perinatal dioxin exposure on neonatal electroencephalography (EEG) activity of the quiet sleep stage in the most contaminated area from Agent Orange in Vietnam. *Int. J. Hyg. Environ. Health* **2021**, *232*, 113661. [CrossRef]
21. Van der Geest, J.N.; Kemner, C.; Verbaten, M.N.; van Engeland, H. Gaze behavior of children with pervasive developmental disorder toward human faces: A fixation time study. *J. Child Psychol. Psychiatry* **2002**, *43*, 669–678. [CrossRef]
22. Rigby, S.N.; Stoesz, B.M.; Jakobson, L.S. Gaze patterns during scene processing in typical adults and adults with autism spectrum disorders. *Res. Autism Spectr. Disord.* **2016**, *25*, 24–36. [CrossRef]
23. Nakano, T.; Tanaka, K.; Endo, Y.; Yamane, Y.; Yamamoto, T.; Nakano, Y.; Ohta, H.; Kato, N.; Kitazawa, S. Atypical gaze patterns in children and adults with autism spectrum disorders dissociated from developmental changes in gaze behaviour. *Proc. Biol Sci.* **2010**, *277*, 2935–2943. [CrossRef]
24. Thao, P.N.; Nishijo, M.; Tai, P.T.; Nghi, T.N.; Quan, L.V.; Anh, T.H.; Vu, P.H.A.; Nishino, Y.; Nishijo, H. Effects of dioxin exposure on gaze behavior in 3-year-old children in Vietnam. *Organohalog. Compd.* **2018**, *80*, 125–128.
25. Fombonne, E.; Rogé, B.; Claverie, J.; Courty, S.; Frémolle, J. Microcephaly and macrocephaly in autism. *J. Autism Dev. Disord.* **1999**, *29*, 113–119. [CrossRef] [PubMed]
26. Gillberg, C.; de Souza, L. Head circumference in autism, Asperger syndrome, and ADHD: A comparative study. *Dev. Med. Child Neurol.* **2002**, *44*, 296–300. [CrossRef]
27. Lainhart, J.E.; Bigler, E.D.; Bocian, M.; Coon, H.; Dinh, E.; Dawson, G.; Deutsch, C.K.; Dunn, M.; Estes, A.; Tager-Flusberg, H.; et al. Head circumference and height in autism: A study by the Collaborative Program of Excellence in Autism. *Am. J. Med. Genet. A* **2006**, *140*, 2257–2574. [CrossRef] [PubMed]
28. Crucitti, J.; Hyde, C.; Enticott, P.G.; Stokes, M.A. Head circumference trends in autism between 0 and 100 months. *Autism* **2020**, *24*, 1726–1739. [CrossRef]
29. The Office of the Vietnam National Steering Committee 33; Hatfield Consultants. *Environmental and Human Health As-Sessment of Dioxin Contamination at Bien Hoa Airbase, Viet Nam*; Final Report; Hatfield Consultants: North Vancouver, BC, Canada, 2011.
30. Schecter, A.; Dai, L.C.; Papke, O.; Prange, J.; Constable, J.D.; Matsuda, M.; Thao, V.D.; Piskac, A.L. Recent dioxin contamination from Agent Orange in residents of a southern Vietnam city. *J. Occup. Environ. Med.* **2001**, *43*, 435–443. [CrossRef]
31. Tawara, K.; Honda, R.; Nishijo, M.; Nakagawa, H. Pretreatment procedure of dioxin analysis for a small volume of human breast milk. *J. Kanazawa Med. Univ.* **2003**, *28*, 17–25.
32. Van den Berg, M.; Birnbaum, L.S.; Denison, M.; De Vito, M.; Farland, W.; Feeley, M.; Fiedler, H.; Hakansson, H.; Hanberg, A.; Haws, L.; et al. The 2005 World Health Organization reevaluation of human and mammalian toxic equivalency factors for dioxins and dioxin-like compounds. *Toxicol. Sci.* **2006**, *93*, 223–241. [CrossRef]
33. Tai, P.T.; Nishijo, M.; Kido, T.; Nakagawa, H.; Maruzeni, S.; Naganuma, R.; Anh, N.T.; Morikawa, Y.; Luong, H.V.; Anh, T.H.; et al. Dioxin concentrations in breast milk of Vietnamese nursing mothers: A survey four decades after the herbicide spraying. *Environ. Sci. Technol.* **2011**, *45*, 6625–6632. [CrossRef] [PubMed]
34. Pham, T.N.; Nishijo, M.; Pham, T.T.; Vu, H.T.; Tran, N.N.; Tran, A.H.; Do, Q.; Takiguchi, T.; Nishino, Y.; Nishijo, H. Dioxin exposure and sexual dimorphism of gaze behavior in prepubertal Vietnamese children living in Da Nang, a hot spot for dioxin contamination. *Sci. Total Environ.* **2020**, *749*, 141083. [CrossRef] [PubMed]
35. Nishijo, M.; Pham, T.T.; Nguyen, A.T.; Tran, N.N.; Nakagawa, H.; Hoang, L.V.; Tran, A.H.; Morikawa, Y.; Ho, M.D.; Kido, T.; et al. 2,3,7,8-Tetrachlorodibenzo-p-dioxin in breast milk increases autistic traits of 3-year-old children in Vietnam. *Mol. Psychiatry* **2014**, *19*, 1220–1226. [CrossRef] [PubMed]
36. Tai, P.T.; Nishijo, M.; Anh, N.T.; Maruzeni, S.; Nakagawa, H.; Van Luong, H.; Anh, T.H.; Honda, R.; Kido, T.; Nishijo, H. Dioxin exposure in breast milk and infant neurodevelopment in Vietnam. *Occup. Environ. Med.* **2013**, *70*, 656–662. [CrossRef] [PubMed]

37. Doi, H.; Nishitani, S.; Fujisawa, T.X.; Nagai, T.; Kakeyama, M.; Maeda, T.; Shinohara, K. Prenatal exposure to a polychlorinated biphenyl (PCB) congener influences fixation duration on biological motion at 4-months-old: A preliminary study. *PLoS ONE* **2013**, *8*, e59196. [CrossRef]
38. Nowack, N.; Wittsiepe, J.; Kasper-Sonnenberg, M.; Wilhelm, M.; Schölmerich, A. Influence of Low-Level Prenatal Exposure to PCDD/Fs and PCBs on Empathizing, Systemizing and Autistic Traits: Results from the Duisburg Birth Cohort Study. *PLoS ONE* **2015**, *10*, e0129906. [CrossRef] [PubMed]
39. Boraston, Z.; Blakemore, S.J. The application of eye-tracking technology in the study of autism. *J. Physiol.* **2007**, *581*, 893–898. [CrossRef]
40. Chakrabarti, B.; Dudbridge, F.; Kent, L.; Wheelwright, S.; Hill-Cawthorne, G.; Allison, C.; Banerjee-Basu, S.; Baron-Cohen, S. Genes related to sex steroids, neural growth, and social-emotional behavior are associated with autistic traits, empathy, and Asperger syndrome. *Autism Res.* **2009**, *2*, 157–177. [CrossRef]
41. Crider, A.; Thakkar, R.; Ahmed, A.O.; Pillai, A. Dysregulation of estrogen receptor beta (ERβ), aromatase (CYP19A1), and ER co-activators in the middle frontal gyrus of autism spectrum disorder subjects. *Mol. Autism* **2014**, *5*, 46. [CrossRef]
42. Dissanayake, C.; Bui, Q.M.; Huggins, R.; Loesch, D.Z. Growth in stature and head circumference in high-functioning autism and Asperger disorder during the first 3 years of life. *Dev. Psychopathol.* **2006**, *18*, 381–393. [CrossRef]
43. Tai, P.T.; Nishijo, M.; Nghi, T.N.; Nakagawa, H.; Van Luong, H.; Anh, T.H.; Nishijo, H. Effects of Perinatal Dioxin Exposure on Development of Children during the First 3 Years of Life. *J. Pediatr.* **2016**, *175*, 159–166. [CrossRef] [PubMed]
44. Tran, N.N.; Pham, T.T.; Ozawa, K.; Nishijo, M.; Nguyen, A.T.; Tran, T.Q.; Hoang, L.V.; Tran, A.H.; Phan, V.H.; Nakai, A.; et al. Impacts of Perinatal Dioxin Exposure on Motor Coordination and Higher Cognitive Development in Vietnamese Preschool Children: A Five-Year Follow-Up. *PLoS ONE* **2016**, *11*, e0147655. [CrossRef] [PubMed]
45. Pham The, T.; Pham Ngoc, T.; Hoang Van, T.; Nishijo, M.; Tran Ngoc, N.; Vu Thi, H.; Hoang Van, L.; Tran Hai, A.; Nishino, Y.; Nishijo, H. Effects of perinatal dioxin exposure on learning abilities of 8-year-old children in Vietnam. *Int. J. Hyg. Environ. Health* **2020**, *223*, 132–141. [CrossRef] [PubMed]
46. Pham-The, T.; Nishijo, M.; Pham-Ngoc, T.; Vu-Thi, H.; Tran-Ngoc, N.; Tran-Hai, A.; Hoang-Van, L.; Nishino, Y.; Nishijo, H. Effects of prenatal dioxin exposure on children behaviors at 8 years of age of age. In Proceedings of the 39th International Symposium on Halogenated Persistent Organic Pollutants—Dioxin in 2019, Kyoto, Japan, 25–30 August 2019; Organizing Committee of the Symposium: Kyoto, Japan, 2019; p. 517.
47. Vu, H.T.; Nishijo, M.; Pham, T.N.; Pham-The, T.; Hoanh, L.V.; Tran, A.H.; Tran, N.N.; Nishino, Y.; Do, Q.; Nishijo, H. Effects of perinatal dioxin exposure on mirror neuron activity in 9-year-old children living in a hot spot of dioxin contamination in Vietnam. *Neuropsychologia* **2021**, *161*, 108001. [CrossRef]
48. Ames, J.; Warner, M.; Siracusa, C.; Signorini, S.; Brambilla, P.; Mocarelli, P.; Eskenazi, B. Prenatal dioxin exposure and neuropsychological functioning in the Seveso Second Generation Health Study. *Int. J. Hyg. Environ. Health* **2019**, *222*, 425–433. [CrossRef]
49. Vreugdenhil, H.J.; Slijper, F.M.; Mulder, P.G.; Weisglas-Kuperus, N. Effects of perinatal exposure to PCBs and dioxins on play behavior in Dutch children at school age. *Environ. Health Perspect.* **2002**, *110*, A593–A598. [CrossRef]
50. Winneke, G.; Ranft, U.; Wittsiepe, J.; Kasper-Sonnenberg, M.; Fürst, P.; Krämer, U.; Seitner, G.; Wilhelm, M. Behavioral sexual dimorphism in school-age children and early developmental exposure to dioxins and PCBs: A follow-up study of the Duisburg Cohort. *Environ. Health Perspect.* **2014**, *122*, 292–298. [CrossRef]
51. Poland, A.; Knutson, J.C. 2,3,7,8-tetrachlorodibenzo-p-dioxin and related halogenated aromatic hydrocarbons: Examination of the mechanism of toxicity. *Annu. Rev. Pharmacol. Toxicol.* **1982**, *22*, 517–554. [CrossRef]
52. Peterson, R.E.; Theobald, H.M.; Kimmel, G.L. Developmental and reproductive toxicity of dioxins and related compounds: Cross-species comparisons. *Crit. Rev. Toxicol.* **1993**, *23*, 283–335. [CrossRef]
53. Takeda, T.; Matsumoto, Y.; Koga, T.; Mutoh, J.; Nishimura, Y.; Shimazoe, T.; Ishii, Y.; Ishida, T.; Yamada, H. Maternal exposure to dioxin disrupts gonadotropin production in fetal rats and imprints defects in sexual behavior. *J. Pharmacol. Exp. Ther.* **2009**, *329*, 1091–1099. [CrossRef]
54. Hattori, Y.; Takeda, T.; Nakamura, A.; Nishida, K.; Shioji, Y.; Fukumitsu, H.; Yamada, H.; Ishii, Y. The aryl hydrocarbon receptor is indispensable for dioxin-induced defects in sexually-dimorphic behaviors due to the reduction in fetal steroidogenesis of the pituitary-gonadal axis in rats. *Biochem. Pharmacol.* **2018**, *154*, 213–221. [CrossRef] [PubMed]
55. Furue, M.; Ishii, Y.; Tsukimori, K.; Tsuji, G. Aryl Hydrocarbon receptor and dioxin-related health hazards-lessons from Yusho. *Int. J. Mol. Sci.* **2021**, *22*, 708. [CrossRef] [PubMed]
56. Dobs, K.; Bülthoff, I.; Schultz, J. Use and Usefulness of Dynamic Face Stimuli for Face Perception Studies-a Review of Behavioral Findings and Methodology. *Front. Psychol.* **2018**, *9*, 1355. [CrossRef] [PubMed]

toxics

MDPI

Article

Alterations in Regional Brain Regional Volume Associated with Dioxin Exposure in Men Living in the Most Dioxin-Contaminated Area in Vietnam: Magnetic Resonance Imaging (MRI) Analysis Using Voxel-Based Morphometry (VBM)

Hoa Thi Vu [1], Thao Ngoc Pham [2], Takashi Yokawa [3], Muneko Nishijo [1], Tai Pham The [2], Quyet Do [2], Yoshikazu Nishino [1] and Hisao Nishijo [4,*]

[1] Department of Public Health, Kanazawa Medical University, Uchinada 920-0293, Japan; vuhoa5593hvqy@gmail.com (H.T.V.); ni-koei@kanazawa-med.ac.jp (M.N.); ynishino@kanazawa-med.ac.jp (Y.N.)

[2] Biomedical and Pharmaceutical Research Center, Vietnamese Military Medical University, Hanoi 10000, Vietnam; phamngocthaovmmu@gmail.com (T.N.P.); taithuy@kanazawa-med.ac.jp (T.P.T.); dobaquyet@yahoo.com (Q.D.)

[3] Kobe BMA Laboratory, BioView Inc., Kobe 650-0047, Japan; yokawa@bioview.co.jp

[4] Faculty of Medicine, University of Toyama, Toyama 930-0194, Japan

* Correspondence: nishijo@med.u-toyama.ac.jp; Tel.: +81-764347215

Citation: Vu, H.T.; Pham, T.N.; Yokawa, T.; Nishijo, M.; The, T.P.; Do, Q.; Nishino, Y.; Nishijo, H. Alterations in Regional Brain Regional Volume Associated with Dioxin Exposure in Men Living in the Most Dioxin-Contaminated Area in Vietnam: Magnetic Resonance Imaging (MRI) Analysis Using Voxel-Based Morphometry (VBM). *Toxics* **2021**, *9*, 353. https://doi.org/10.3390/toxics9120353

Academic Editor: Jaymie R. Meliker

Received: 5 November 2021
Accepted: 13 December 2021
Published: 15 December 2021

Abstract: To clarify the influence of dioxin exposure on brain morphometry, the present study investigated associations between dioxin exposure at high levels and brain structural irregularities in 32 Vietnamese men. Two exposure markers were used: blood dioxin levels, as a marker of exposure in adulthood, and perinatal dioxin exposure, estimated by maternal residency in a dioxin-contaminated area during pregnancy. All subjects underwent brain magnetic resonance imaging (MRI) scans. We analyzed correlations between regional gray matter volumes and blood dioxin levels, and compared regional volumes between men with and without perinatal dioxin exposure using the voxel-based morphometry (VBM) tool from Statistical Parametric Mapping 12 (SPM12). Blood 2,3,7,8-tetrachlorodibenzo-p-dioxin (TCDD) was associated with low volume of the medial temporal pole and fusiform gyrus. Toxic equivalency (TEQ)-PCDDs were correlated with low medial temporal pole volume. However, 1,2,3,4,7,8-HxCDD was associated with high middle frontal gyrus and cerebellum volume. In men with perinatal dioxin exposure, the left inferior frontal gyrus pars orbitalis volume was significantly lower than in those without perinatal exposure. These results suggest that dioxin exposure during the perinatal period and in adulthood may alter regional brain volume, which might lead to cognitive deficits and unusual social emotional behavior in Vietnamese men living in dioxin-contaminated areas.

Keywords: dioxin; neuro imaging analysis; brain regional volume; adults; Vietnam

1. Introduction

During "Operation Ranch Hand" between 1961 and 1971, the US Armed Forces sprayed large quantities of herbicides, such as Agent Orange, causing widespread dioxin contamination in Southern Vietnam. Nearly 50 years later, Dwernychuk (2005) reported that dioxin residues remained in both the environment and in humans residing in the sprayed areas of Vietnam, particularly around several former US airbases located in Bien Hoa, Da Nang, and Phu Cat [1]. Bien Hoa airbase is recognized as the largest hotspot of dioxin contamination in Vietnam; the Office of the Vietnam National Steering Committee 33 and Hatfield Consultants (2011) reported the detection of large quantities of 2,3,7,8-tetrachlorodibenzo-p-dioxin (TCDD), as high as 61,400 pg/g dry weight in soil, originating from Agent Orange, in samples from the Bien Hoa airbase [2]. Blood dioxin concentrations in men living around the Phu Cat and Bien Hoa airbases were four to five times higher than those in an unsprayed area in Northern Vietnam [3]. TCDD concentrations in blood samples of military workers in Bien Hoa airbase were two to five times higher than those in people working in Da Nang and Phu Cat airbases, respectively [4].

Our previous Vietnam-based study reported that perinatal exposure to toxic equivalency values of polychlorodibenzodioxins/furans (TEQ-PCDD/Fs) was associated with poor neurodevelopmental scores for all domains, and that TCDD exposure was associated with increased autistic traits in 3-year-old children in an area around Da Nang airbase [5]. In the areas around Bien Hoa airbase, where TCDD levels in breast milk were two times higher, but TEQ-PCDD/Fs levels were two-thirds of those in Da Nang; we found that boys with high exposure to TCDD and other PCDD congeners showed lower expressive and composite language scores and lower gross motor scores than those in a low exposure group [6]. In 2015, neonatal electroencephalography (EEG) was measured in a group of neonates recruited in the same area, and perinatal dioxin exposure was found to be associated with altered EEG power and coherence, poor neurodevelopment, and poor gaze behavior at 2 years of age [7,8]. These findings suggest that dioxin exposure may influence development of the fetal brain, leading to neurodevelopment disorders such as autism spectrum disorder (ASD) in later life.

ASD is a neurodevelopmental disorder that presents with social communication deficits and restricted repetitive patterns of behavior, interests, and activities. Voxel-based morphometry (VBM) is a technique applied to brain MRI to investigate focal differences in brain anatomy. Several VBM studies have demonstrated structural brain irregularities in patients with ASD [9–11]. These studies raised the possibility that children with autistic traits caused by dioxin exposure might have different brain structures to children not exposed to dioxin. However, the MRI examination of children is difficult because of the requirement for sedation. Therefore, in the present study, we recruited the fathers of children in our birth cohort recruited in Bien Hoa in 2015, and used VBM to investigate associations between dioxin exposure and brain structural irregularities in men living near Bien Hoa airbase.

2. Materials and Methods

2.1. Study Subjects

In 2018, 55 fathers of children living in communes around Bien Hoa airbase whose neonatal EEGs were examined in 2015 were invited to join a survey to investigate their dioxin exposure levels according to blood analysis, and 40 of them responded (73% participant rates). In 2019, these 40 men whose blood dioxin levels were measured were invited to brain MRI examinations, but only 33 men (60%) participated in the present study. Seven men were too busy with their works or families to join the study on the examination days. Moreover, one participant was considered an outlier because of a very high TCDD blood level (371.5 pg/g lipid) and was excluded from the data analysis. Consequently, 32 men were included in the final data analysis of the present study.

As regional brain volumes in adulthood may be programmed during the fetal period [12], we investigated associations between the men's regional brain volumes and perinatal dioxin exposure estimated from their mothers' residential information before their births. The subjects were interviewed to retrieve details of their mothers' residency before they were born, and 12 mothers (37.5%) were found to have lived in Bien Hoa during pregnancy (1970 to 1992) with one of our subjects, suggesting that these men might have been exposed to a high level of TCDD during the fetal period. Schecter et al. (2001) reported very high levels of TCDD in breast milk; 133–1832 (pg/g lipid) in samples collected in 1970–1973 and 2.1–11 (pg/g lipid) in 1985–1988 among Bien Hoa residents, although the levels were decreased over years [13]. They also reported high blood levels of TCDD, 2.0–164 (pg/g lipid) in samples collected in 1999 [13]. Furthermore, it is supposed that all mothers fed babies breast milk before weaning, since infant formula was not readily available and not a common feeding method for residents in Bien Hoa after the long wartime. These findings suggest that the mothers of our subjects as well as the subjects, who had lived in Bien Hoa before and after birth, were exposed to TCDD originating from Agent Orange even after spraying herbicide was terminated.

Written informed consent was obtained from all men according to a process reviewed and approved by the Health Departments of Bien Hoa City and Dong Nai Prefecture. The institutional ethics board for medical and health research involving human subjects at Kanazawa Medical University approved the study design (Approval Code: No. I-424, Approval Date: 19 September 2017).

2.2. Dioxin Measurements in Whole Blood

Around twenty milliliters of venous blood were collected by nursing staff at Bien Hoa health center. Whole blood samples were frozen and transferred to Kanazawa Medical University in Japan, and the levels of 17 2,3,7,8-substituted PCDD/Fs (7 congeners of PCDDs and 10 congeners of polychlorinated dibenzofurans; PCDFs) and four non-ortho-polychlorinated biphenyl (PCB) congeners (3,3,4,4-tetra-chlorinated biphenyl [TCB#77]; 3,4,4,5-tetra-chlorinated biphenyl [TCB#81]; 3,3,4,4,5-penta-chlorinated biphenyl [PeCB#126]; and 3,3,4,4,5,5-hexa-chlorinated biphenyl [HxCB#169]) were measured. Whole blood samples were frozen and dehydrated in an EYELA freeze dryer (FDU-1200; Tokyo-rika Inc., Tokyo, Japan), and fat was extracted using an ASE-200 accelerated solvent extractor (Dionex Corporation, Sunnyvale, CA, USA). Then, 13C-labeled 2,3,7,8-substituted PCDDs/Fs (DF-LCS-A40; Wellington Inc., Ontario, Canada) and 13C-labeled dioxin-like (dl)-PCBs (DLPCB-CL-A20; Kanto Chemical Co., Inc., Tokyo, Japan) were added to samples as internal standards. After chromatography using a multilayered silica gel column to purify samples, a single-layered column of activated carbon was used to separate and collect the PCDD/Fs and non-ortho PCBs fraction. The final extracted solution was concentrated by evaporation, and 17 PCDD/F congeners and 4 non-ortho PCB congeners were quantified using a gas chromatograph (HP-6980; Hewlett-Packard, Palo Alto, CA, USA) equipped with a high-resolution mass spectrometer (MStation-JMS700; JEOL, Tokyo, Japan). The limit of detection (LOD) for each congener of PCDDs/Fs/ non-ortho PCBs with numbers of samples lower than LODs were shown in Table 1. For quantification of each congener, measurement values were used if they were higher than LODs, but, if not (below LODs), halves of the LODs were used as a quantification values.

Table 1. Dioxin and non-ortho-PCB levels in blood (n = 32).

Dioxin Congeners PCDD Congeners (pg/g Lipid)	LOD (ppt)	Below LOD No.	Below LOD %	GM	GSD	Min	Max
2,3,7,8-TCDD	0.03	2	6.3	6.4	2.1	1.5	56.2
1,2,3,7,8-PeCDD	0.03	0	0	10.6	1.5	4.4	22.4
1,2,3,4,7,8-HxCDD	0.03	2	6.3	6.1	1.4	3.0	11.7
1,2,3,6,7,8-HxCDD	0.02	0	0	13.3	1.5	5.5	37.2
1,2,3,7,8,9-HxCDD	0.03	0	0	7.2	1.6	3.1	31.6
1,2,3,4,6,7,8-HpCDD	0.04	0	0	36.2	1.8	12.0	295
OctaCDD	0.04	0	0	1027	1.6	490	2570
PCDF congeners (pg/g lipid)							
2,3,7,8-TCDF	0.02	2	6.3	4.4	1.6	1.6	8.3
1,2,3,7,8-PeCDF	0.03	7	21.9	4.5	1.7	1.3	10.2
2,3,4,7,8-PeCDF	0.04	0	0	13.6	1.3	6.6	21.9
1,2,3,4,7,8-HxCDF	0.02	0	0	14.9	1.4	7.2	30.2
1,2,3,6,7,8-HxCDF	0.02	0	0	10.8	1.5	2.9	19.1
1,2,3,7,8,9-HxCDF	0.02	13	40.6	3.8	2.0	0.9	67.6
2,3,4,6,7,8-HxCDF	0.02	1	3.1	3.8	1.6	1.3	9.8
1,2,3,4,6,7,8-HpCDF	0.02	0	0	13.6	1.6	5.2	50.1
1,2,3,4,7,8,9-HpCDF	0.04	10	31.3	4.8	1.8	1.1	15.1
OctaCDF	0.07	15	46.9	10.0	1.7	2.2	21.4

Table 1. *Cont.*

Dioxin Congeners PCDD Congeners (pg/g Lipid)	LOD (ppt)	Below LOD		GM	GSD	Min	Max
		No.	%				
Nonortho-PCB (pg/g lipid)							
TCB #77	0.05	0	0	56.8	1.6	17.8	186
TCB #81	0.05	31	96.9	ND	ND	ND	ND
PeCB #126	0.28	30	93.8	ND	ND	ND	ND
HxCB #169	0.05	2	6.3	43.6	1.7	16.2	145
TEQs (pg-TEQ/g lipid)							
TEQ-PCDDs				21.8	1.5	9.1	69.2
TEQ-PCDFs				8.6	1.3	5.0	14.1
TEQ-PCDDs/Fs				30.8	1.4	14.5	74.1
TEQ-nonortho PCBs				1.1	3.8	0.01	5.89
TEQ-PCDDs/Fs/nonorthoPCBs				32.4	1.4	14.5	79.4

n: number of subjects; GM: geometrical mean; GSD: geometrical standard deviation; Min: minimum; Max: maximum. LOD (ppt): limit of detection (ppt: parts per trillion): it was defined as a signal-to-noise (S/N) ratio of peak height of chromato-gram = 3. Average of LODs in measurements of each dioxin congener is shown. ND: not detected.

The TEQ values of PCDDs, PCDFs, and non-ortho PCBs in each sample were calculated by multiplying each congener concentration by its TEQ factor referenced from the World Health Organization 2005 TEQ factors list [14]. The geometrical means and ranges of 17 PCDD/F congeners, TCB#77, and HxCB#169, and the TEQs for PCDDs, PCDFs, PCDDs/Fs, non-ortho PCBs, and PCDDs/Fs/non-ortho PCBs in the blood after lipid base calculations are shown in Table 1. Because of levels below detection limits for almost all samples (>93%), the values of TCB#81 and PeCB#126 congeners are not shown in Table 1.

2.3. MRI Data Acquisition and Voxel-Based Morphometry

All subjects underwent brain MRI scans on a Siemens Magnetom Trio Tim system 3T-scanner (Siemens, Erlangen, Germany) at the Department of Diagnostic Imaging in Dong Nai General Hospital, Vietnam. High-resolution T1-weighted images with good contrast between gray and white matter were collected. The image parameters included: TR = 1520 ms, TE = 2.07 ms, flip angle = 9°, slice thickness = 0.9 mm, 192 slices, and field of view =230 × 230 mm.

All images subjected to voxel-based morphometry (VBM) analysis were preprocessed using the Computational Anatomy Toolbox (CAT12, vCAT12.7-RC1; Structural Brain Mapping group, Jena University Hospital, Jena, Germany; http://dbm.neuro.uni-jena.de/cat/ (accessed on 14 December 2021)) in the Statistical Parametric Mapping 12 software package (SPM12; The Wellcome Centre for Human Neuroimaging, London, UK; https://www.fil.ion.ucl.ac.uk/spm/software/spm12/ (accessed on 14 December 2021)) running under MATLAB software (The Mathworks, Inc., Natick, MA, USA). The image preprocessing steps performed were based on a standard protocol (http://www.neuro.uni-jena.de/cat12/CAT12-Manual.pdf (accessed on 14 December 2021)). All T1-weighted images were corrected for rough bias, affine registered to a template image in MNI152 space, then segmented into gray matter, white matter, and cerebrospinal fluid maps using the segmentation tools in SPM12 [15]. The segmented images were spatially normalized to the same total brain volume using the DARTEL algorithm [16] and smoothed with an isotropic Gaussian kernel (full-width half-maximum = 8 mm).

The total gray matter, white matter, cerebrospinal fluid, and intracranial volumes were estimated using CAT12. The total brain volume was calculated from the sum of the volume of gray and white matter [17].

2.4. Statistical Analysis

The correlations of VBM-derived regional brain volumes with blood dioxin levels were evaluated using a multiple regression model in SPM12 using the data of the 32 men.

The normalization process of the DARTEL algorithm normalized the global brain volume of each subject to the same value while conserving regional differences in brain matter volume. Blood dioxin congener levels were treated as a covariate of interest after base-10 logarithm transformation to improve normality. The total intracranial brain volume and age of the subjects were treated as confounding covariates. Two linear contrasts (positive or negative correlations) were used to estimate associations between blood dioxin levels and brain regions.

Two-sample *t*-tests in SPM12 were used to compare regional brain volumes between subjects with and without perinatal dioxin exposure indicated by maternal residency in Bien Hoa during pregnancy. These tests were performed after adjusting for total intracranial brain volume and age.

The Montreal Neurological Institute (MNI) coordinates of the voxel of maximal statistical significance in each cluster (connected voxels defined by a prespecified statistical threshold) were extracted. For all VBM analyses, the cluster-based false-discovery rate (FDR) for multiple comparisons combined with the peak detection threshold was used for testing statistical significance [18]. The Anatomy toolbox in SPM12 [19] was used to identify anatomical regions showing significant results in the VBM analysis. $p < 0.05$ was considered statistically significant in all tests.

SPSS version 22.0 (IBM, Armonk, NY, USA) was used for statistical analyses of global brain volume. General linear models were used to compare global brain volume parameters (independent variables) between subjects with and without perinatal dioxin exposure (a fixed factor), after adjusting for confounding factors correlated with global brain volume (covariates), including age (years) and height (cm).

3. Results

3.1. Relevant Factors Associated with Brain Volume and Dioxin Exposure

The characteristics of the subjects are shown in Table 2. The mean (±standard deviation (SD)) age and educational years of the subjects were 35.5 ± 5.9 and 11.9 ± 3.1 years, respectively. Twenty-two men (68.8% of the total subjects) graduated high school and/or schools with upper levels. Twenty-six men (81.3%) consumed alcohol, but only four men (12.5%) took alcohol every day. Their mean body mass index (BMI) was 24.2 and 14 men (43.8%) were obese (BMI ≥ 25). Most subjects were right-handed (87.5%). The mean (±SD) residential period near to Bien Hoa airbase was 22.1 ± 14.6 years. Five men (15.6%) were mechanics who worked on machines that belonged to the airbase, and four men were soldiers. Ten men (31.3%) used herbicides and pesticides in the growing of vegetables in their gardens.

The dioxin and non-ortho PCB concentrations in the blood were compared according to occupation, consumption of food grown in the airbase, use of herbicides and pesticides, and length of residency. The concentrations of TCDD and TEQs of PCDDs and PCDDs/Fs/non-orthoPCBs were significantly higher in people who worked in the airbase than in those with other jobs (t = −3.27, p = 0.003; t = −3.41, p = 0.002; t = −3.28, p = 0.003, respectively), and in subjects using herbicides and pesticides than in those not using them (t = −2.41, p = 0.022; t = −2.69, p = 0.012; t = −2.29, p = 0.030, respectively). However, these differences were not statistically significant after adjusting confounding factors. The geometric mean TCDD concentration of two men who consumed dioxin-contaminated food grown in the airbase was four times higher than that of other subjects. There were no significant correlations between the concentrations of any of the PCDDs/Fs/non-orthoPCB congeners and the length of residency.

Table 2. Characteristics of the subjects (n = 32).

Characteristics	Mean, [n]	SD, (%)	Min	Max
Age (years)	35.5	5.9	25.3	49
Education (years)	11.9	3.1	1	16
Smoking	[15]	(46.9)		
Alcohol consumption	[26]	(81.3)		
Weight (kilogram)	66.3	9.4	46	81.2
Height (cm)	165.5	5.0	154	178
BMI	24.2	3.0	17.1	28.2
Right hand dominant	[28]	(87.5)		
Length of residency (years)	22.1	14.6	1	44
Used herbicides	[10]	(31.3)		
Worked nearby industrial park	[15]	(46.9)		
Job (% of jobs related to the airbase)	[5]	(15.6)		
Their mothers lived in Bien Hoa during pregnancy	[12]	(37.5)		

n: number of subjects; SD: standard deviation; BMI: body mass index.

3.2. *Associations between Current Dioxin Exposure (Indicated by Blood Dioxins) and Brain Volume*

3.2.1. Global Brain Volume Analysis

We analyzed simple correlations (Spearman's ρ) between global brain volume and levels of 17 congeners of PCDD/Fs and 2 congeners of non-ortho-PCBs (data not shown). Global gray matter volume (cm^3) was significantly inversely correlated with 1,2,3,4,7,8-Hexachlorodibenzo-p-dioxin (HxCDD) ($\rho = -0.378$, $p = 0.033$), but positively correlated with Octachlorodibenzo-p-dioxin ($\rho = 0.369$, $p = 0.038$) and 1,2,3,4,6,7,8-Heptachlorodibenzo furan ($\rho = 0.354$, $p = 0.047$). A significant positive correlation was also found between global cerebrospinal fluid volume (cm^3) and HxCB#169 ($\rho = 0.446$, $p = 0.011$). However, these correlations between global gray matter and dioxin and PCB congeners were not significant after adjusting for confounding factors such as age and height (data not shown). No significant correlations with any dioxin congeners in the blood were found for total white matter volume, total brain volume, and total intracranial volume, even before adjusting for confounding factors.

3.2.2. VBM Analyses; Brain Regions in Which Gray Matter Volume Correlated with Blood Dioxin Levels

Regions with low gray matter volume (inverse correlations) and high gray matter volume (positive correlations) in association with high dioxin exposure are shown in Table 3 and Figures 1–3. Figure 1 shows an example of inverse correlation between the voxel values of the left fusiform gyrus at [−27 (x), 8 (y), −47 (z)] and blood TCDD levels (see below for the details).

Table 3. Brain regions significantly correlated with blood levels of dioxin congeners in men after adjusting for total intracranial volume and age (FDR-corrected at $p < 0.05$).

Dioxin Congeners	Brain Regions	No of Voxels in Each Cluster (k)	Peak Z Scores	MNI Coordinates		
				x	y	z
Inverse correlations						
TCDD	Anterior temporal cortex	905				
	(Left medial temporal pole)		3.81	−41	20	−38
	(Left fusiform gyrus)		3.90	−27	8	−47
TEQ-PCDDs	Left medial temporal pole	333	3.63	−39	21	−38
Positive correlations						
1,2,3,4,7,8-HxCDD	Left cerebellum lobule VII	373	3.87	−42	−60	−57
	Right middle frontal gyrus	505	3.86	41	6	60

FDR = false discovery rate; MNI = Montreal Neurological Institute.

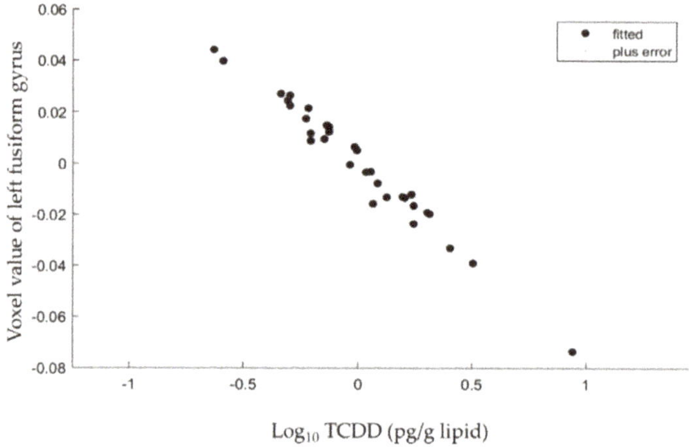

Figure 1. Relationships between the voxel values of the left fusiform gyrus at [−27 (x), 8 (y), −47 (z)] and blood TCDD levels in the 32 subjects. There was a significant and inverse correlation between the two parameters after adjusting for total intracranial volume and age of the subjects.

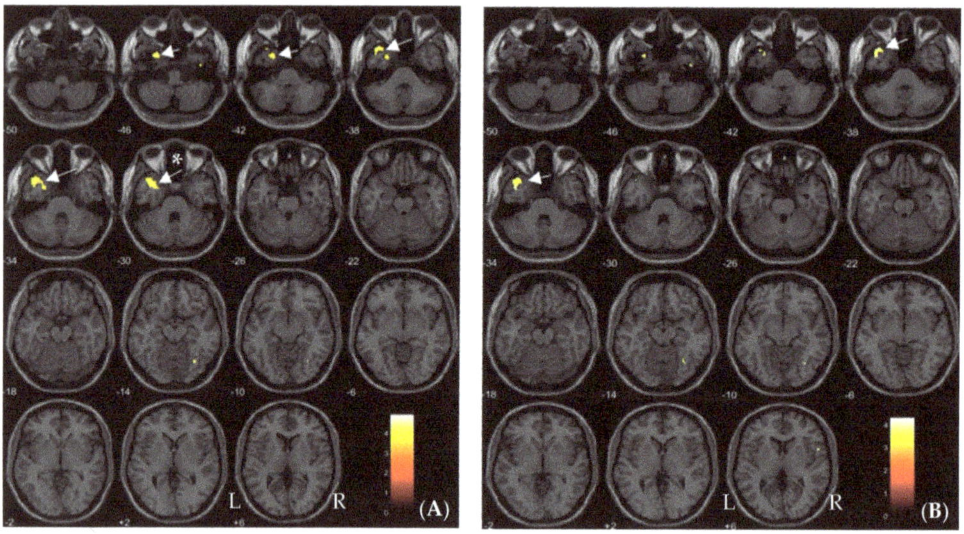

Figure 2. Brain regions showing significant negative correlations with blood levels of TCDD (**A**) and TEQ-PCDDs (**B**) (FDR-corrected at $p < 0.05$) are indicated by yellow color and arrows on axial MRI. (**A**) shows the correlations in the left medial temporal pole (arrows without *) and fusiform gyrus (an arrow with *) and (**B**) shows those in the left medial temporal pole (arrows). L and R indicate the left and right sides of the hemispheres, respectively. Each value below each brain slice indicates each value in MNI z-coordinates.

Volume in the anterior temporal cortex including the left medial temporal pole and fusiform gyrus showed significant inverse correlations with TCDD ($p < 0.05$, FDR-corrected; Table 3, Figure 2A). The left medial temporal pole was significantly inversely correlated with TEQ-PCDDs ($p < 0.05$, FDR-corrected; Table 3, Figure 2B). In contrast, the left cerebellum and right middle frontal gyrus volume were significantly positively correlated with 1,2,3,4,7,8-HxCDD exposure ($p < 0.05$, FDR-corrected; Table 3, Figure 3).

(A)

(B)

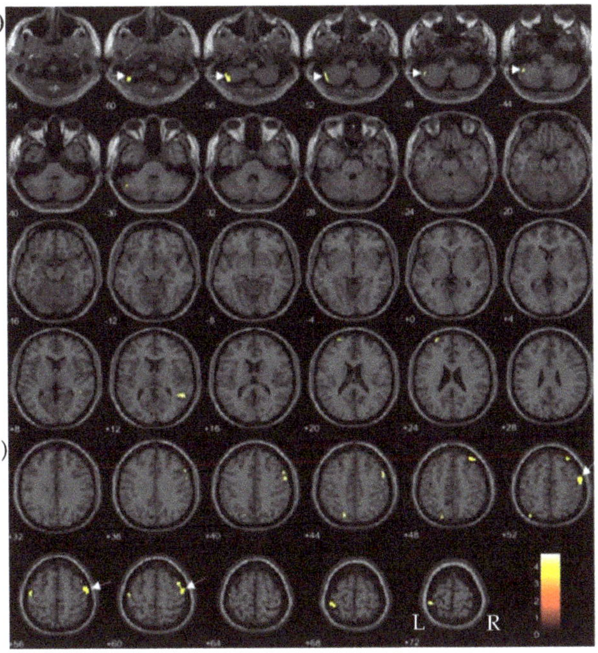

Figure 3. Brain regions positively correlated with 1,2,3,4,7,8-HexaCDD (FDR-corrected at $p < 0.05$) are indicated by yellow color and arrows on axial MRI. (**A**) shows the correlations in the left cerebellum lobule VII (arrows) and (**B**) shows those in the right middle frontal gyrus (arrows). L and R indicate the left and right sides of the hemispheres, respectively. Each value below each brain slice indicates each value in MNI z-coordinates.

3.3. Comparisons of Global and Regional Brain Volumes between Men with and without Possible Perinatal Dioxin Exposure

3.3.1. Global Brain Volume Analyses

Total gray matter volume was significantly higher in men supposed to have been subject to perinatal dioxin exposure than in men without perinatal exposure ($p = 0.005$; effect size (ES) = 0.252; Table 4). However, no significant differences in white matter volume and total cerebrospinal fluid were observed between men with and without perinatal dioxin exposure. Total brain volume and total intracranial volume were significantly higher in men with perinatal exposure than in men without exposure ($p = 0.020$, ES = 0.178 for total brain volume; $p = 0.034$, ES = 0.151 for total intracranial volume; Table 4).

Table 4. Adjusted comparisons of global brain volumes between men with and without perinatal dioxin exposure estimated according to their mothers' residency in Bien Hoa during pregnancy.

Perinatal Dioxin Exposure	Without (n = 20)			With (n = 12)				
		95%CI			95%CI			
Global Volumes	Adj Mean	Lower	Upper	Adj Mean	Lower	Upper	p-Value	ES
Gray matter (GM) (cm^3)	615	601	629	651	632	670	0.005	0.252
White matter (WM) (cm^3)	527	508	545	551	527	576	0.118	0.085
Cerebrospinal fluid (CSF) (cm^3)	300	283	317	306	284	329	0.651	0.007
Total brain volume (TBV) (cm^3)	1142	1111	1172	1202	1163	1242	0.020	0.178
Total intracranial volume (TIV) (cm^3)	1443	1401	1484	1517	1463	1570	0.034	0.151

n: number of subjects; adj mean: adjusted mean; CI: confidence interval; ES: effect size. Covariates: age (years) and height (cm).

3.3.2. VBM Analysis; Brain Regions in Which Gray Matter Volume Differed between Men with and without Estimated Perinatal Dioxin Exposure

The gray matter volume of the left inferior frontal gyrus (IFG) pars orbitalis was significantly lower in men with perinatal dioxin exposure than in with men without perinatal exposure after adjusting for total intracranial volume and age ($p < 0.05$, FDR-corrected; Table 5, Figure 4). However, no area showed a higher volume in men with perinatal exposure in comparison with men without exposure (Table 5).

Table 5. Brain regions showing significant contrasts in volume between men with and without perinatal dioxin exposure after adjusting for total intracranial volume and age (FDR-corrected at $p < 0.05$).

Brain Regions	No of Voxels in Each Cluster (k)	Peak Z Scores	MNI Coordinates x	y	z
Without exposure > With exposure					
Left inferior frontal gyrus pars orbitalis	414	3.86	−32	39	−23
With exposure > Without exposure					
No brain region	-	-	-	-	-

FDR = false discovery rate; MNI = Montreal Neurological Institute. Perinatal exposure was estimated on their mothers' residency in Bien Hoa during pregnancy. "-" indicates that no significant correlations were detected.

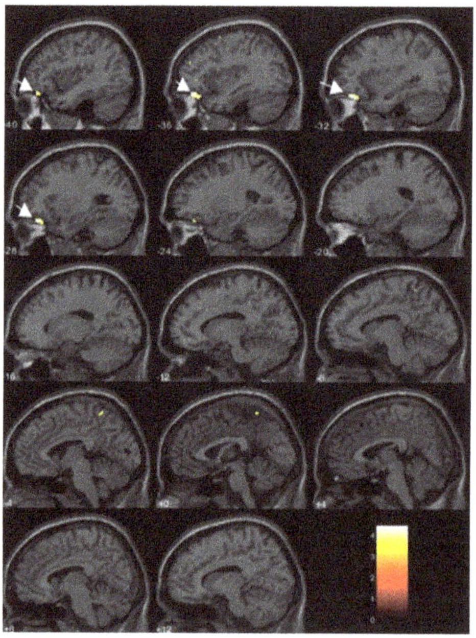

Figure 4. Brain regions associated with perinatal dioxin exposure (the left inferior frontal gyrus pars orbitalis indicated by yellow color and arrows on sagittal MRI slices). The volume was significantly lower in men with estimated perinatal dioxin exposure than in men without perinatal exposure (FDR-corrected at $p < 0.05$). Each value below each brain slice indicates each value in MNI x-coordinates.

4. Discussion

4.1. Relationships between Blood Dioxin Levels and Relevant Factors

In the present study, the geometric means of TCDD in the blood of the present subjects were 2.5-fold and 4.3-fold higher than those in the Phu Cat hotspot of dioxin contamination and a non-polluted area in Vietnam, respectively [20]. In particular, the blood TCDD levels were significantly higher in soldiers and mechanics who worked on machines belonging to the airbase than in those with other occupations. Similarly, Van Manh et al.

(2021) reported TCDD concentrations in the blood of military workers in three Vietnamese airbases exposed to dioxins and found the highest concentrations in those who worked in Bien Hoa airbase [4]. These results suggest that a job related to an airbase is an important relevant factor for high levels of TCDD in men.

Consumption of foods such as vegetables and fish grown within the Bien Hoa airbase was also a significant factor relevant to high TCDD levels. Although the growing and catching of vegetables and fish within the airbase is forbidden, some people still consume vegetables and fish grown within it, thereby increasing the dioxin burden on their body. In addition, the use of herbicides and pesticides for agriculture in and around Bien Hoa airbase could be also associated with high levels of TCDD in the blood of the present subjects. However, agricultural chemicals produced in recent years should not be contaminated with TCDD, and should not therefore be a source of TCDD exposure. The enhanced levels in those working with herbicides and pesticides may be a factor co-incident with working with TCDD-contaminated soil in their farming areas, possibly from exposure to dust containing contaminated soil particles. These results suggest that TCDD originating from Agent Orange still remains a source of exposure to dioxin contamination in Bien Hoa airbase, and that TCDD in blood may reflect exposure levels in adulthood.

4.2. Regional Brain Volume Changes Associated with Blood Dioxin Levels

We showed that high TCDD and TEQ-PCDD levels in the blood were significantly associated with low gray matter volume in the left medial temporal pole and fusiform gyrus. Conversely, high 1,2,3,4,7,8-HxCDD exposure was associated with gray matter volume in the left cerebellum lobule VII and the right middle frontal gyrus. However, no significant correlations between blood dioxin levels and global volumes were found. These results suggest that dioxin exposure in adulthood might affect regional brain structure to alter brain volumes.

Previously, a follow-up study performed on workers in the Czech Republic 35 years after exposure to high levels of TCDD reported alteration of EEG and visual evoked potentials, and reduction of perfusion in various locations in the brain [21]. However, the effects of dioxin exposure on alterations to regional brain volumes are unknown. To our knowledge, this is the first study reporting morphological brain alterations associated with dioxin exposure. We found that opposing effects were associated with high dioxin levels, observing both volume increases and volume decreases. Previous animal studies suggested that the effects of persistent organic pollutants (POPs), including dioxins, might be non-monotonic: POPs exert both promoting and suppressing effects on neuronal dendritic growth, depending on their dose and the brain region, which may lead to brain volume increases and decreases, respectively [22–24].

We found that the gray matter volume in the fusiform gyrus was significantly inversely correlated with blood TCDD level. The fusiform gyrus is a large region in the temporal cortex implicated in high-level visual functions such as face perception, object recognition, and reading [25]. Grecucci et al. (2016) reported that some brain regions such as the middle frontal gyrus, fusiform gyrus, and cerebellum exhibited functional and structural abnormalities in patients with ASD in comparison with controls, referring to these regions as an autism-specific structural network [26]. The cerebellum is active in cognitive functions (including language) and executive functions [27], which might be disturbed in ASD. Another finding of our study is the association between low gray matter volume in the medial temporal pole and high exposure to TCDD and TEQ-PCDDs. The medial temporal pole may be involved in olfactory processing and connects with the orbitofrontal cortex and other emotion-related areas [28]. These results indicated that dioxin exposure in adulthood, including exposure to TCDD and 1,2,3,4,7,8-HxCDD, induced brain structural alterations in the areas involved in cognitive and emotional functions. Further following-up studies of our subjects with an assessment of cognitive and emotional functions are required to determine if these structural changes induce cognitive and emotional impairments.

4.3. Global and Regional Brain Volume Alterations Associated with Estimated Perinatal Dioxin Exposure

Our results indicate that perinatal dioxin exposure, which was estimated on the basis of the mother's residency in Bien Hoa during pregnancy, was significantly associated with increased global gray matter volume, total brain volume, and total intracranial volume. To our knowledge, no previous study has investigated the effects of perinatal dioxin exposure on structural brain changes in humans. Although some previous animal studies reported reduced cortical thickness in several brain areas and changes in cortical cell numbers and cell distributions in TCDD-exposed rats [29,30], recent animal studies suggest that dioxins in low doses may lead to neuronal dendritic overgrowth [22,24], which may in turn lead to increases in brain volume.

Our previous study found that perinatal TCDD exposure increased autistic traits in Vietnamese children [5]. It was also reported that Vietnamese children perinatally exposed to dioxin showed disturbed mirror neuron activity [31], which is one of the social cognitive deficits present in autism. Moreover, extensive studies have reported brain enlargement in children [32–34], adolescents, and adults with autism [35,36]. In a VBM study of 833 children and adults, Riddle et al. (2017) recognized higher total brain volume and gray matter volume of approximately 1–2% in the ASD group compared with the typically-developing group [11]. Consistently, an animal model of autism showed dendritic overgrowth [37]. These similar results (increased brain volume) suggest that perinatal dioxin exposure and autism may share a similar pathogenesis: dendritic overgrowth could be a cause of pathogenesis resulting in brain enlargement and cognitive deficits.

Additionally, to identify regional gray matter abnormalities in men subjected to perinatal dioxin exposure, we applied VBM analysis. After adjusting for total intracranial volume and age, this analysis showed regional gray matter volume reduction in the left IFG pars orbitalis compared with a group without such exposure. The IFG pars orbitalis is important in the comprehension of emotional signals and semantic communication in humans [38]. Salmond et al. (2003) reported abnormality of orbitofrontal cortical volume in adolescents with ASD [39]. Furthermore, we also reported that dioxin exposure was associated with attention deficit hyperactivity disorder (ADHD) traits in Vietnamese children [40]. Several studies in ADHD children also demonstrated reduced gray matter volume in the left orbitofrontal cortex relative to healthy children [41,42]. Volume reduction of the left IFG pars orbitalis may be associated with symptoms of inattention and/or impulsivity in ADHD [42]. Taking these findings together, perinatal dioxin exposure may increase total gray matter volume as well as total brain volume, and may also decrease gray matter volume in the IFG pars orbitalis, suggesting that perinatal dioxin exposure affects cognitive and social functions by altering global and local brain development, leading to adverse neurodevelopment and traits, such as ASD and ADHD.

4.4. Limitations

One limitation of our study is the use of the residency in Bien Hoa of the subjects' mothers to estimate perinatal dioxin exposure. A previous study reported high levels of TCDD in breast milk in Bien Hoa residents (see Materials and Methods, Section 2.1), suggesting that our subjects were exposed to dioxins during the perinatal period. Furthermore, our previous studies reported that dioxin levels in breast milk were associated with neurodevelopment of children [5–8]. The present results indicated that the residency in Bien Hoa of the subjects' mothers, which suggests that the subjects were subjected to perinatal exposure to dioxins, was significantly associated with changes in gray matter volumes that were not associated with the adulthood blood dioxin levels of the subjects. These findings suggest possible effects of perinatal exposure to dioxins on brain morphometry although dose-effect relationships are unknown. We previously reported adverse effects of perinatal dioxin exposure on the neurodevelopment of Vietnamese children. However, MRI examination of children is difficult in Vietnam because of the requirement for sedation. In the future, we will perform MRI studies in these subjects when they become adolescents, to investigate the presence of brain structural alterations associated with perinatal dioxin exposure. The small sample size is also another limitation of the present study.

We also presented regional structural alterations in the brain associated with blood dioxin levels that were different from those associated with perinatal exposure, suggesting the neurotoxic effects of dioxin on the developed brain in adulthood. However, we did not examine the subjects' cognitive functions and emotional behavior in the present study. In future, we will investigate psychiatric symptoms and behavior in residents of Bien Hoa to clarify the effects of dioxins on the mental health of adults without perinatal dioxin exposure.

5. Conclusions

The present results demonstrated that the residency in Bien Hoa of the subjects' mothers was significantly associated with changes in gray matter volumes (i.e., increased global gray matter volume as well as decreased gray matter volume in the left IFG pars orbitalis), which was not associated with the present blood dioxin levels of the subjects. This finding suggests the possible effects of perinatal exposure to dioxins on gray matter volumes. Dioxin exposure in adulthood, indicated by blood levels, was associated with low gray matter volume in the fusiform gyrus and the medial temporal pole, but high gray matter volume in the middle frontal gyrus and cerebellum. These results suggest that dioxin exposure during the perinatal period and in adulthood is associated with regional brain alterations, which might lead to cognitive deficits and unusual social emotional behavior (increased autistic traits) in Vietnamese men living in a hot spot of dioxin contamination.

Author Contributions: Conceptualization, M.N., Y.N. and H.N.; methodology, T.P.T., T.N.P. and M.N.; software, T.Y.; investigation, T.P.T., T.N.P. and M.N.; data curation, H.T.V.; writing—original draft preparation, H.T.V. and M.N.; writing—review and editing, H.N.; project administration, Q.D.; funding acquisition, M.N. All authors have read and agreed to the published version of the manuscript.

Funding: This work was supported partly by the Ministry of Education, Sports, Science and Culture, Japan, Grant-in-Aid for Scientific Research (17H04665 and 18K19709). These funders played no role in the study design, data collection and analysis, decision to publish, or preparation of the manuscript.

Institutional Review Board Statement: The study was conducted according to the guidelines of the Declaration of Helsinki and approved by the Institutional Review Boards of the Health Department of Bien Hoa City, Dong Nai prefecture and the Kanazawa Medical University (No. I424).

Informed Consent Statement: Informed consent was obtained from all subjects involved in the study.

Data Availability Statement: The data presented in this study are available on request to the corresponding author. The data are not publicly available due to the personal information (MRI data).

Acknowledgments: We would like to thank all fathers participating in this study. We are grateful to Phan Huy Anh Vu in Dong Nai Health Department, Nguyen Xuan Hung in Bien Hoa Health Center, medical staff in communes around Bien Hoa airbase, and Le Thi Phuong Tram and medical staff in the Diagnostic Imaging Department in Dong Nai general hospital for their collaboration. We thank Karl Embleton, from Edanz (https://jp.edanz.com/ (accessed on 14 December 2021)), for editing a draft of this manuscript.

Conflicts of Interest: The authors declare no conflict of interest.

References

1. Dwernychuk, L.W. Dioxin hot spots in Vietnam. *Chemosphere* **2005**, *60*, 998–999. [CrossRef]
2. The Office of the Vietnam National Steering Committee 33; Hatfield Consultants. *Environmental and Human Health Assessment of Dioxin Contamination at Bien Hoa Airbase, Viet Nam*; Final Report; Hatfield Consultants: North Vancouver, BC, Canada, 2011.
3. Van Luong, H.; Tai, P.T.; Nishijo, M.; Trung, D.M.; Thao, P.N.; Van Son, P.; Van Long, N.; Linh, N.T.; Nishijo, H. Association of dioxin exposure and reproductive hormone levels in men living near the Bien Hoa airbase, Vietnam. *Sci. Total Environ.* **2018**, *628–629*, 484–489. [CrossRef] [PubMed]

4. Van Manh, P.; Tai, P.T.; Phuong, N.M.; Nishijo, M.; Trung, D.M.; Thao, P.N.; Son, H.A.; Van Tuan, T.; Van Chuyen, N.; Van Long, N.; et al. Serum dioxin concentrations in military workers at three dioxin-contaminated airbases in Vietnam. *Chemosphere* **2021**, *266*, 129024. [CrossRef]
5. Nishijo, M.; Pham, T.T.; Nguyen, A.T.; Tran, N.N.; Nakagawa, H.; Hoang, L.V.; Tran, A.H.; Morikawa, Y.; Ho, M.D.; Kido, T.; et al. 2,3,7,8-Tetrachlorodibenzo-p-dioxin in breast milk increases autistic traits of 3-year-old children in Vietnam. *Mol. Psychiatry* **2014**, *19*, 1220–1226. [CrossRef] [PubMed]
6. Pham, N.T.; Nishijo, M.; Pham, T.T.; Tran, N.N.; Le, V.Q.; Tran, H.A.; Phan, H.A.V.; Nishino, Y.; Nishijo, H. Perinatal dioxin exposure and neurodevelopment of 2-year-old Vietnamese children in the most contaminated area from Agent Orange in Vietnam. *Sci. Total Environ.* **2019**, *678*, 217–226. [CrossRef]
7. Nghiem, G.T.; Nishijo, M.; Pham, T.N.; Ito, M.; Pham, T.T.; Tran, A.H.; Nishimaru, H.; Nishino, Y.; Nishijo, H. Adverse effects of maternal dioxin exposure on fetal brain development before birth assessed by neonatal electroencephalography (EEG) leading to poor neurodevelopment; a 2-year follow-up study. *Sci. Total Environ.* **2019**, *667*, 718–729. [CrossRef]
8. Pham, N.T.; Nishijo, M.; Nghiem, T.T.G.; Pham, T.T.; Tran, N.N.; Le, V.Q.; Vu, T.H.; Tran, H.A.; Phan, H.A.V.; Do, Q.; et al. Effects of perinatal dioxin exposure on neonatal electroencephalography (EEG) activity of the quiet sleep stage in the most contaminated area from Agent Orange in Vietnam. *Int. J. Hyg. Environ. Health* **2021**, *232*, 113661. [CrossRef] [PubMed]
9. Albajara Saenz, A.; Van Schuerbeek, P.; Baijot, S.; Septier, M.; Deconinck, N.; Defresne, P.; Delvenne, V.; Passeri, G.; Raeymaekers, H.; Slama, H.; et al. Disorder-specific brain volumetric abnormalities in Attention-Deficit/Hyperactivity Disorder relative to Autism Spectrum Disorder. *PLoS ONE* **2020**, *15*, e0241856.
10. Lim, L.; Chantiluke, K.; Cubillo, A.I.; Smith, A.B.; Simmons, A.; Mehta, M.A.; Rubia, K. Disorder-specific grey matter deficits in attention deficit hyperactivity disorder relative to autism spectrum disorder. *Psychol. Med.* **2015**, *45*, 965–976. [CrossRef]
11. Riddle, K.; Cascio, C.J.; Woodward, N.D. Brain structure in autism: A voxel-based morphometry analysis of the Autism Brain Imaging Database Exchange (ABIDE). *Brain Imaging Behav.* **2017**, *11*, 541–551. [CrossRef] [PubMed]
12. Raznahan, A.; Greenstein, D.; Lee, N.R.; Clasen, L.S.; Giedd, J.N. Prenatal growth in humans and postnatal brain maturation into late adolescence. *Proc. Natl. Acad. Sci. USA* **2012**, *109*, 11366–11371. [CrossRef] [PubMed]
13. Schecter, A.; Dai, L.C.; Papke, O.; Prange, J.; Constable, J.D.; Matsuda, M.; Thao, V.D.; Piskac, A.L. Recent dioxin contamination from Agent Orange in residents of a southern Vietnam city. *J. Occup. Environ. Med.* **2001**, *43*, 435–443. [CrossRef]
14. Van den Berg, M.; Birnbaum, L.S.; Denison, M.; De Vito, M.; Farland, W.; Feeley, M.; Fiedler, H.; Hakansson, H.; Hanberg, A.; Haws, L.; et al. The 2005 World Health Organization Reevaluation of Human and Mammalian Toxic Equivalency Factors for Dioxins and Dioxin-Like Compounds. *Toxicol. Sci.* **2006**, *93*, 223–241. [CrossRef] [PubMed]
15. Ashburner, J.; Friston, K.J. Unified segmentation. *NeuroImage* **2005**, *26*, 839–851. [CrossRef] [PubMed]
16. Ashburner, J. A fast diffeomorphic image registration algorithm. *NeuroImage* **2007**, *38*, 95–113. [CrossRef] [PubMed]
17. Nordenskjold, R.; Malmberg, F.; Larsson, E.M.; Simmons, A.; Brooks, S.J.; Lind, L.; Ahlstrom, H.; Johansson, L.; Kullberg, J. Intracranial volume estimated with commonly used methods could introduce bias in studies including brain volume measurements. *NeuroImage* **2013**, *83*, 355–360. [CrossRef]
18. Genovese, C.R.; Lazar, N.A.; Nichols, T. Thresholding of statistical maps in functional neuroimaging using the false discovery rate. *NeuroImage* **2002**, *15*, 870–878. [CrossRef]
19. Eickhoff, S.B.; Stephan, K.E.; Mohlberg, H.; Grefkes, C.; Fink, G.R.; Amunts, K.; Zilles, K. A new SPM toolbox for combining probabilistic cytoarchitectonic maps and functional imaging data. *NeuroImage* **2005**, *25*, 1325–1335. [CrossRef] [PubMed]
20. Manh, H.D.; Kido, T.; Okamoto, R.; Xianliang, S.; Anhle, T.; Supratman, S.; Maruzeni, S.; Nishijo, M.; Nakagawa, H.; Honma, S.; et al. Serum dioxin levels in Vietnamese men more than 40 years after herbicide spraying. *Environ. Sci. Technol.* **2014**, *48*, 3496–3503. [CrossRef]
21. Urban, P.; Pelclova, D.; Lukas, E.; Kupka, K.; Preiss, J.; Fenclova, Z.; Smerhovsky, Z. Neurological and neurophysiological examinations on workers with chronic poisoning by 2,3,7,8-TCDD: Follow-up 35 years after exposure. *Eur. J. Neurol.* **2007**, *14*, 213–218. [CrossRef]
22. Kimura, E.; Kubo, K.; Matsuyoshi, C.; Benner, S.; Hosokawa, M.; Endo, T.; Ling, W.; Kohda, M.; Yokoyama, K.; Nakajima, K.; et al. Developmental origin of abnormal dendritic growth in the mouse brain induced by in utero disruption of aryl hydrocarbon receptor signaling. *Neurotoxicol. Teratol.* **2015**, *52*, 42–50. [CrossRef]
23. Gileadi, T.E.; Swamy, A.K.; Hore, Z.; Horswell, S.; Ellegood, J.; Mohan, C.; Mizuno, K.; Lundebye, A.K.; Giese, K.P.; Stockinger, B.; et al. Effects of Low-Dose Gestational TCDD Exposure on Behavior and on Hippocampal Neuron Morphology and Gene Expression in Mice. *Environ. Health Perspect.* **2021**, *129*, 57002. [CrossRef]
24. Latchney, S.E.; Majewska, A.K. Persistent organic pollutants at the synapse: Shared phenotypes and converging mechanisms of developmental neurotoxicity. *Dev. Neurobiol.* **2021**, *81*, 623–652. [CrossRef]
25. Weiner, K.S.; Zilles, K. The anatomical and functional specialization of the fusiform gyrus. *Neuropsychologia* **2016**, *83*, 48–62. [CrossRef] [PubMed]
26. Grecucci, A.; Rubicondo, D.; Siugzdaite, R.; Surian, L.; Job, R. Uncovering the Social Deficits in the Autistic Brain. A Source-Based Morphometric Study. *Front. Neurosci.* **2016**, *10*, 388. [CrossRef] [PubMed]
27. Becker, E.B.; Stoodley, C.J. Autism spectrum disorder and the cerebellum. *Int. Rev. Neurobiol.* **2013**, *113*, 1–34.
28. Fan, L.; Wang, J.; Zhang, Y.; Han, W.; Yu, C.; Jiang, T. Connectivity-based parcellation of the human temporal pole using diffusion tensor imaging. *Cereb. Cortex* **2014**, *24*, 3365–3378. [CrossRef] [PubMed]

29. Zareba, G.; Hojo, R.; Zareba, K.M.; Watanabe, C.; Markowski, V.P.; Baggs, R.B.; Weiss, B. Sexually dimorphic alterations of brain cortical dominance in rats prenatally exposed to TCDD. *J. Appl. Toxicol.* **2002**, *22*, 129–137. [CrossRef] [PubMed]
30. Hojo, R.; Zareba, G.; Kai, J.W.; Baggs, R.B.; Weiss, B. Sex-specific alterations of cerebral cortical cell size in rats exposed prenatally to dioxin. *J. Appl. Toxicol.* **2006**, *26*, 25–34. [CrossRef]
31. Vu, H.T.; Nishijo, M.; Pham, T.N.; Pham-The, T.; Hoanh, L.V.; Tran, A.H.; Tran, N.N.; Nishino, Y.; Do, Q.; Nishijo, H. Effects of perinatal dioxin exposure on mirror neuron activity in 9-year-old children living in a hot spot of dioxin contamination in Vietnam. *Neuropsychologia* **2021**, *161*, 108001. [CrossRef]
32. Sparks, B.F.; Friedman, S.D.; Shaw, D.W.; Aylward, E.H.; Echelard, D.; Artru, A.A.; Maravilla, K.R.; Giedd, J.N.; Munson, J.; Dawson, G.; et al. Brain structural abnormalities in young children with autism spectrum disorder. *Neurology* **2002**, *59*, 184–192. [CrossRef]
33. Palmen, S.J.; Hulshoff Pol, H.E.; Kemner, C.; Schnack, H.G.; Durston, S.; Lahuis, B.E.; Kahn, R.S.; Van Engeland, H. Increased gray-matter volume in medication-naive high-functioning children with autism spectrum disorder. *Psychol. Med.* **2005**, *35*, 561–570. [CrossRef]
34. Stanfield, A.C.; McIntosh, A.M.; Spencer, M.D.; Philip, R.; Gaur, S.; Lawrie, S.M. Towards a neuroanatomy of autism: A systematic review and meta-analysis of structural magnetic resonance imaging studies. *Eur. Psychiatry* **2008**, *23*, 289–299. [CrossRef] [PubMed]
35. Freitag, C.M.; Luders, E.; Hulst, H.E.; Narr, K.L.; Thompson, P.M.; Toga, A.W.; Krick, C.; Konrad, C. Total brain volume and corpus callosum size in medication-naive adolescents and young adults with autism spectrum disorder. *Biol. Psychiatry* **2009**, *66*, 316–319. [CrossRef] [PubMed]
36. Hazlett, H.C.; Poe, M.D.; Gerig, G.; Smith, R.G.; Piven, J. Cortical gray and white brain tissue volume in adolescents and adults with autism. *Biol. Psychiatry* **2006**, *59*, 1–6. [CrossRef]
37. Cheng, N.; Alshammari, F.; Hughes, E.; Khanbabaei, M.; Rho, J.M. Dendritic overgrowth and elevated ERK signaling during neonatal development in a mouse model of autism. *PLoS ONE* **2017**, *12*, e0179409. [CrossRef] [PubMed]
38. Belyk, M.; Brown, S.; Lim, J.; Kotz, S.A. Convergence of semantics and emotional expression within the IFG pars orbitalis. *NeuroImage* **2017**, *156*, 240–248. [CrossRef] [PubMed]
39. Salmond, C.H.; de Haan, M.; Friston, K.J.; Gadian, D.G.; Vargha-Khadem, F. Investigating individual differences in brain abnormalities in autism. *Philos. Trans. R. Soc. Lond. B Biol. Sci.* **2003**, *358*, 405–413. [CrossRef]
40. Pham-The, T.; Nishijo, M.; Pham-Ngoc, T.; Vu-Thi, H.; Tran-Ngoc, N.; Tran-Hai, A.; Hoang-Van, L.; Nishino, Y.; Nishijo, H. Effects of prenatal dioxin exposure on children behaviors at 8 years of age of age. In Proceedings of the 39th International Symposium on Halogenated Persistent Organic Pollutants—Dioxin in 2019, Kyoto, Japan, 25–30 August 2019; p. 517.
41. Kumar, U.; Arya, A.; Agarwal, V. Neural alterations in ADHD children as indicated by voxel-based cortical thickness and morphometry analysis. *Brain Dev.* **2017**, *39*, 403–410. [CrossRef]
42. Nickel, K.; Tebartz van Elst, L.; Manko, J.; Unterrainer, J.; Rauh, R.; Klein, C.; Endres, D.; Kaller, C.P.; Mader, I.; Riedel, A.; et al. Inferior Frontal Gyrus Volume Loss Distinguishes Between Autism and (Comorbid) Attention-Deficit/Hyperactivity Disorder—A FreeSurfer Analysis in Children. *Front. Psychiatry* **2018**, *9*, 521. [CrossRef]

toxics

MDPI

Review

Neurodevelopmental Effects of Perinatal TCDD Exposure Differ from Those of Other PCDD/Fs in Vietnamese Children Living near the Former US Air Base in Da Nang, Vietnam

Nghi Ngoc Tran [1], Tai Pham-The [2], Thao Ngoc Pham [2], Hoa Thi Vu [3], Khue Ngoc Luong [1] and Muneko Nishijo [3,*]

1. Ministry of Health, Vietnam Government, Hanoi 10000, Vietnam
2. Faculty of Medicine, Vietnam Military Medical University, Hanoi 10000, Vietnam
3. Department of Public Health, Kanazawa Medial University, Uchinada 252-0815, Ishikawa, Japan

* Correspondence: ni-koei@kanazawa-med.ac.jp; Tel.: +81-76-218-8093

Abstract: This study reports that children exposed to 2,3,7,8-tetra-chlorodibenzo-p-dioxin (TCDD), the major toxin in Agent Orange, from the breast milk of mothers residing near the former Da Nang US air base in Vietnam may have specific alterations in higher brain functions, resulting in social and communication deficits, including autism spectrum disorder (ASD). After the age of 8 years, girls with high TCDD showed increased attention deficit hyperactivity disorder (ADHD)-like behaviors and altered mirror neuron activity, which is often observed in children with ASD. However, no significant relationship between autistic traits and toxic equivalency values of polychlorinated dibenzodioxins and polychlorinated dibenzofurans (TEQ-PCDD/Fs) was found in these children. Notably, boys with high levels of TEQ-PCDD/Fs showed poor language and motor development in the first 3 years of life, although boys with high TCDD levels did not. However, at 8 years of age, boys with high TCDD showed reading learning difficulties, a neurodevelopmental disorder. These findings suggest that perinatal TCDD exposure impacts social–emotional cognitive functions, leading to sex-specific neurodevelopmental disorders—learning difficulty in boys and ADHD in girls. Future studies with a greater number of children exposed to high levels of TCDD are necessary to estimate the threshold values for neurodevelopmental effects.

Keywords: dioxins; 2,3,7,8-tetrachlorodibenzo-p-dioxin (TCDD); neurodevelopment; birth cohort study; children; Vietnam

Citation: Tran, N.N.; Pham-The, T.; Pham, T.N.; Vu, H.T.; Luong, K.N.; Nishijo, M. Neurodevelopmental Effects of Perinatal TCDD Exposure Differ from Those of Other PCDD/Fs in Vietnamese Children Living near the Former US Air Base in Da Nang, Vietnam. *Toxics* **2023**, *11*, 103. https://doi.org/10.3390/toxics11020103

Academic Editor: Małgorzata Dobrzyńska

Received: 26 November 2022
Revised: 16 January 2023
Accepted: 18 January 2023
Published: 21 January 2023

1. Introduction

Da Nang Air Base, located in central Vietnam, is a former U.S. air base contaminated with dioxins from the use of Agent Orange and other herbicides containing 2,3,7,8-tetrachlorodibenzo-p-dioxin (TCDD) during the Vietnam War. We previously measured levels of 17 polychlorinated dibenzodioxins and polychlorinated dibenzofuran (PCDD/F) congeners in the breast milk of mothers residing nearby Da Nang Air Base and found these to be three- to four-fold higher than those in the breast milk of mothers living in unsprayed areas [1]. These results suggested that environmental contamination by dioxins is still high enough to increase health risks in the residents living in hot spots of dioxin contamination, even 40 years after the end of the war.

The effects of dioxins on infant neurodevelopment have previously been examined in studies in Europe [2] and Japan [3]. We therefore followed up infants whose mother's milk samples were examined in our survey in Da Nang [1]. Our investigation of this Da Nang birth cohort, from 4 months to 8 years of age, identified adverse effects of dioxin exposure on infant and child neurodevelopment in several age groups, using standardized test batteries and parent rating scales for different aspects of neurodevelopment [4–12].

In the follow-up study of the Da Nang cohort, performed when the children reached 3 years of age, we found increased autistic traits (poor social and communication abilities, such as is observed in autism spectrum disorder (ASD)) in children of both sexes, associated with increased TCDD exposure but not TEQ-PCDD/F exposure [4]. Interestingly, we found that only boys showed poor language and motor development associated with high TEQ-PCDD/F exposure (but not TCDD exposure). These results suggest that TCDD may have specific neurodevelopmental effects that differ from those of other dioxin congeners.

In boys at 5 years of age, cognitive ability, assessed with the non-verbal index (NVI) of the Kaufman Assessment Battery for Children, second edition (KABC-II), was significantly lower in the high TCDD group, whereas poor coordination movement skills, indicated by the Movement Assessment Buttery for Children, second edition (MABC-2), were observed in those with high TEQ-PCDD/F levels [9]. At the same age, girls in the high TCDD group showed higher unusual behavioral scores, indicating increased autistic behavior associated with TCDD exposure, but not TEQ-PCDD/F exposure [10]. Moreover, at 8 years of age, boys with high TCDD showed reading learning disabilities, lower language achievement test scores, and poor reading skills [11].

Here, we reviewed the results of long-term follow-up studies of birth cohorts to clarify the specific neurotoxic effects of TCDD in perinatally exposed children. We focused on the following three aspects of toxicity: (1) the stage of neurodevelopment affected by TCDD; (2) differences between sexes regarding sensitivity to TCDD; and (3) the threshold values of TCDD for significant neurodevelopmental impairments.

2. Materials and Methods

2.1. Literature Searches

At first, a search of the literature was conducted using PubMed in English to find birth cohorts followed up for a long time to investigate associations between perinatal PCDD/F exposure and neurodevelopment. The following terms were used in the search procedure: "dioxins" AND "neurodevelopment" AND "birth cohort". We found three birth cohorts followed up for a long period of time until school age, including the Hokkaido birth cohort in Japan, the Duisburg birth cohort in Germany, and the Da Nang birth cohort in Vietnam.

The TCDD exposure levels in maternal blood were too low (0.9 pg/g lipid of the geometrical mean (GM) and 3.1 pg/g lipid of maximum) and examined within narrow ranges (the 25th value is under the detection limit and the 75th value is 1.4 pg/g lipid) [3], and therefore, no significant associations between neurodevelopment and TCDD were obtained in the follow-up studies of the Hokkaido birth cohort at 6 months [3,13], 18 months [14], 42 months [15], or 13 years of age [16]. In the follow-up studies of the Duisburg birth cohort, the authors used only TEQ-PCDD/Fs as dioxin exposure markers (13.55 of pg TEQ/g lipid of the GM and 12.45–14.75 pg TEQ/g lipid at 95% confidence intervals in blood samples) [17] and showed no association between TCDD and behavioral indices in children at 2 years [18], 6–8 years [19], or 9–10 years [17,20]. Only the Da Nang birth cohort studies that showed associations between neurodevelopmental markers and TCDD and other PCDD/F congeners in their follow-up studies until 8 years of age [4–12] were thus selected for the present review.

2.2. Profile of the Da Nang Birth Cohort

The Da Nang cohort in Vietnam consisted of 241 mother–infant pairs (137 boys and 104 girls) living in the dioxin hot spots in the Thanh Khe and Son Tra districts in Da Nang City, located within 10 km of Da Nang Air Base [1]. These mother–infant pairs were recruited for the study by obstetricians at each district hospital when admitted for delivery. The criteria for recruitment were as follows: (1) mothers who resided in the study districts for at least the duration of their pregnancy; (2) babies who were full-term and healthy at birth; and (3) mothers who had no complications during pregnancy and childbirth.

As perinatal exposure markers, the dioxin levels in the maternal breast milk were used in all follow-up studies. A breast milk sample was collected from each nursing mother 1 month after birth with the assistance of a midwife or medical worker. Approximately 10 mL of breast milk from each sample was used to quantify the levels of 17 different 2,3,7,8-

substituted PCDD and PCDF congeners by the established method of analysis described in detail elsewhere [1]. The toxic equivalent factors for calculating the toxic equivalents (TEQ) of PCDDs/Fs (TEQ-PCDDs/Fs) were referenced from the WHO 2005-TEF [21]. The subjects were divided into two to four groups according to the levels of TCDD and TEQ-PCDD/Fs, with cut-off values calculated using the GM and geometrical standard deviation (GSD) of dioxin levels in the breast milk of 138 nursing mothers in unsprayed areas, as follows: GM $\times$ GSD^3 of the TCDD level (3.5 pg/g lipid) for the high TCDD group and GM $\times$ GSD^4 of the TEQ-PCDD/F level for the high TEQ-PCDD/Fs group (17.6 pg TEQ/g lipid).

3. Results

Nine articles published by our group on neurodevelopment in infants and children of various ages from the Da Nang cohort [4–12] are presented, with the age and sex of the subjects and evaluation methods in Table 1 for studies from 4 months to 3 years of age and in Table 2 for studies from 5 to 8 years of age.

Table 1. Neurodevelopment and perinatal dioxin exposure in infants and children aged 4 months to 3 years.

Authors	Age	Neurodevelopmental Markers (Test Battery/Scale)	Results
Tai (Pham-The) et al. [5]	4 months *N* = 210	Cognitive, language, motor scale scores (Bayley III)	Infants with high TEQ-PCDD/Fs (≥17.6) showed lower cognitive and fine motor scores. Infants with moderately higher TCDD (1.8–3.5) showed lower scores for all domains. The increased scores in infants with TCDD ≥ 3.5 were not significant.
Nishijo M et al. [6]	4 months *N* = 210	Cognitive, language, motor scale scores (Bayley III)	Only boys showed lower expressive language scores in those with high TEQ-PCDD/Fs (≥17.6).
Pham TT (Pham-The) et al. [7]	1 year *N* = 214	Cognitive, language, motor, social–emotional, adaptive behavior scale scores (Bayley III)	The high TCDD group (≥3.5) and high TEQ-PCDD/Fs group (≥17.6) showed lower social–emotional scale scores. However, no difference in other developmental scale scores was found among the other groups.
Nishijo M et al. [4]	3 years *N* = 198	Cognitive, language, motor, social–emotional, adaptive behavior scale scales (Bayley III)	Boys with high TEQ-PCDD/Fs (≥17.6) showed lower scores in all domains, except the fine motor scale, compared to the lower group (<17.6). However, no difference was observed in girls.
		Total score (TOT), DSM-IV-TR Scale (DSM), social communication (SC), unusual behavior (UB) scores (ASRS)	In both sexes, those with high TCDD (≥3.5) showed high TOT and DSM scores (increased autistic traits) compared to the low exposure group.
Tai (Pham-The) et al. [8]	4 months to 3 years *N* = 217	Cognitive, language, motor scale scores (Bayley III)	Among boys, the high TCDD group (≥3.5) showed lower marginal means of composite motor scores and gross motor scores. The high TEQ-PCDD/Fs (≥17.6) group showed lower marginal means of expressive language scores. However, no differences in any scales were found among girls.

Units: pg/g lipid for TCDD and pg TEQ/g lipid for TEQ-PCDD/Fs, *N*: number of subjects, Bayley-III: Bayley Scales of Infant and Toddler Development, 3rd edition; ASRS: Autism Spectrum Rating Scale.

Table 2. Neurodevelopment and perinatal dioxin exposure in children aged 5–8 years.

Authors	Age	Neurodevelopmental Markers (Test Battery/Scale)	Results
Tran NN et al. [9]	5 years, *N* = 181	Non-verbal index for cognitive functions (NVI), sequence/general ability of short-term memory (Seq/GSM), simultaneous/general ability of visual processing (Sim/GV) scores (KABC-II)	Only boys with high TCDD (≥2.5) showed lower scores in the NVI and pattern reasoning scores, an NVI component, compared to boys with lower TCDD (<2.5). However, no association with TEQ-PCDD/Fs was found in either sex.
		Total scale for coordination movement skills (TOTAL), manual dexterity (MD), aiming and catching (A&C), balance (BAL) scores (MABC-2)	In boys only, the high TEQ-PCDD/Fs group (≥17.6) showed lower TOTAL and BAL scores compared to the low (<11.5) and moderate (11.5–17.6) exposure groups. However, no association with TCDD was found in either sex.
Pham-The et al. [10]	5 years, *N* = 163	Total scale of ADHD symptoms (ADHD), inattention scale (Inattention), impulsivity and hyperactivity scale (Hyperactivity) scores (ADHD-RS)	Boys with high TCDD (≥ 3.0) showed higher hyperactivity and ADHD scores. However, no association was found with TEQ-PCDD/Fs (≥ 17.6) in boys and with either dioxin marker in girls.
		Total score (TOT), DSM-IV-TR Scale (DSM), social communication (SC), unusual behavior (UB) scores (ASRS)	In girls, those with high TCDD (≥3.0) showed higher UB scores. However, there was no association between any ASRS index and exposure markers in boys.
	8 years, *N* = 163	Total scale of ADHD symptoms (ADHD), inattention scale (Inattention), impulsivity and hyperactivity scale (Hyperactivity) scores (ADHD-RS)	Girls with high TCDD (≥3.0) or high TEQ-PCDD/Fs (>17.6) showed higher hyperactivity scores. However, no difference in any ADHD scores between high and low exposure groups in boys.
Pham-The et al. [11]	8 years, *N* = 185	CLDQ reading, CLDQ math scores (CLDQ)	In boys only, those with higher TCDD (≥3.5) and higher 1,2,3,4,6,7,8-HpCDD (>10.0) showed higher CLDQ reading scores compared to the lower exposure groups.
		Vietnamese and mathematics test scores (Achievement tests)	In boys, the high TCDD (≥3.5) group showed low language achievement scores and the high 1,2,3,4,7,8-HxCDD group showed lower mathematics and language scores. Those with high 1,2,3,4,6,7,8-HpCDD showed lower mathematics scores.
		Speed and number of errors for reading a short Vietnamese passage (Reading test)	In boys, high TCDD (≥3.5) group showed higher reading errors. The high TEQ-PCDD/Fs group (≥17.6) showed higher reading errors. The high 1,2,3,4,7,8-HxCDD and 1,2,3,4,6,7,8-HpCDD groups showed lower reading speed and higher reading errors.
Thao P.N. et al. [12]	8 years, *N* = 172	Duration of fixed gaze behavior on the picture (Eye tracking)	High TCDD exposure (≥3.5) increased the feminine index for viewing human line drawings in girls. Boys with high TEQ-PCDD/Fs (≥17.6) displayed a high feminine index. Almost all PCDD congeners were associated with an increased index in boys.

Units: pg/g lipid for TCDD and pg TEQ/g lipid for TEQ-PCDD/Fs, ASRS: Autism Spectrum Rating Scale; ADHD-RS: Attention Deficit Hyperactivity Disorder Rating Scale, CLDQ: Colorado Learning Difficulty Questionnaire.

To evaluate the general neurodevelopmental status of infants and children, standardized test batteries suitable for their age were used in each study (Tables 1 and 2). To assess cognitive abilities, language, and motor development in infants and children, the Bayley Scales of Infant and Toddler Development, third edition (Bayley III) was used when the infants were approximately 4 months, 1 year, and 3 years of age. At 5 years of age, cognitive ability was assessed with the KABC-II, and coordination motor ability was examined with the MABC-2. For children aged 8 years, school performance related to their learning ability was evaluated using achievement tests for mathematics and Vietnamese, and with a reading test of a passage of a Vietnamese story.

Parent rating scales for several neurodevelopmental disorders were also used in these studies. To assess behaviors associated with autism spectrum disorder (ASD), the Autism Spectrum Rating Scale (ASRS; Multi Health Systems Inc., North Tonawanda, NY, USA) was used for children 3 and 5 years of age. In 5- and 8-year-old children, symptoms of attention deficit hyperactivity disorder (ADHD) were evaluated using the ADHD Rating Scale (ADHD-RS). The Colorado Learning Difficulties Questionnaire (CLDQ) [22] was used to assess the risk for learning difficulty at 8 years of age.

3.1. Associations between Dioxin Exposure Markers and Neurodevelopment in Children Aged 4 Months to 3 Years of Age (Table 1)

In infants 4 months of age in the Da Nang cohort, the fine motor scores were significantly lower in the high TEQ-PCDD/Fs exposure group (≥17.6 pg TEQ/g lipid). Increased levels of 1,2,3,7,8-pentachloro-dibenzo-p-dioxin (1,2,3,7,8-PeCDD), 1,2,3,7,8,9-hexa-chlorodibenzo-p-dioxin (HxCDD), and 1,2,3,4,6,7,8-hepta-chlorodibenzo-p-dioxin (1,2,3,4,6,7,8-HpCDD), which contributed to the elevated TEQ-PCDD/F levels, were also significantly associated with reduced fine motor scores, suggesting that PCDD congeners other than TCDD may influence infant neurodevelopment [5]. In a stratified analysis according to infant gender at the same age, only boys with high TEQ-PCDD/Fs (≥17.6 pg TEQ/g lipid) showed significantly lower Bayley III scores for expressive language development [6]. However, no increased Bayley III scores were observed in the high TCDD group (≥3.5 pg/g lipid) in either sex.

When the Da Nang birth cohort reached 1 year of age, we examined their neurodevelopment using the Bayley III scale again. However, no significant alterations in cognition, language, or motor scale scores associated with TEQ-PCDD/F levels were found in children of either sex, whereas lower social–emotional scale scores were observed in both the high TCDD group (≥3.5 pg/g lipid) and the high PCDD/Fs group (≥17.6 pg TEQ/g lipid) [7].

At 3 years of age, boys with higher TEQ-PCDD/Fs (≥17.6 pg TEQ/g lipid) showed lower scores in all domains, except for the fine motor scale, compared to the low-exposure group (<17.6). However, no difference was observed in girls [4]. The high TCDD group (≥3.5 pg/g lipid) showed high total and DSM-IV-TR Scale (DSM) scores in the ASRS, suggesting increased autistic traits, compared to the low exposure group.

Tai et al. (2016) [8] examined children using the Bayley III scale at 4 months, 1 year, and 3 years of age and analyzed all results longitudinally. These investigators found that boys with high TEQ-PCDD/Fs (≥17.6 pg TEQ/g lipid) showed lower marginal means of expressive language scores, and boys with high TCDD (≥3.5 pg/g lipid) showed lower marginal means of motor scores, particularly gross motor scores. However, no difference in any neurodevelopmental score associated with TEQ-PCDD/F or TCDD level was found in girls.

Taken together, the findings suggest that perinatal exposure to TEQ-PCDD/Fs of ≥17.6 pg TEQ/g lipid may influence neurodevelopment, particularly in boys in the first 3 years of life, although affected ability and skills may differ according to age. Furthermore, TCDD exposure of ≥3.5 pg/g lipid may increase autistic traits in both sexes, but its effect on general neurodevelopment may be limited and differ from the effects of other PCDD/F congeners.

3.2. Associations between Dioxin Exposure Markers and Neurodevelopment in Children Aged 5–8 Years of Age (Table 2)

At 5 years of age, only boys showed low non-verbal index (NVI) scores in the KABC-II, indicating cognitive deficit in the high TCDD group (≥2.5 pg/g lipid) [9]. Increased ADHD Rating Scale scores without increased ASRS scores were found in the high TCDD group (≥3.0 pg/g lipid) in boys [10]. In contrast, girls in the high TCDD group showed higher unusual behavior scores, an ASRS subscale score, although no increase in the ADHD Rating Scale score was found [10]. In addition, in boys with high TEQ-PCDD/Fs (≥17.6 pg TEQ/g lipid), MABC-2 scores, an index of coordination motor skills, were significantly lower; however, the NVI, ADHD, and ASRS rating scale scores were not associated with TEQ-PCDD/Fs levels [9,10].

At 8 years of age, boys with high TCDD (≥3.5 pg/g lipid) showed increased reading scores in the CLDQ, indicating a reading learning disability, such as dyslexia. They also showed significantly lower language achievement scores and higher reading errors in the reading test compared to boys with lower TCDD [1]. Boys with high 1,2,3,4,6,7,8-HpCDD (≥10.0 pg/g lipid) also showed higher CLDQ reading scores (more difficult), lower reading speed, and higher reading errors. Boys with high 1,2,3,4,7,8-HxCDD showed lower reading speed and higher reading errors, while the high TEQ-PCDD/Fs group (≥17.6 pg TEQ/g lipid) displayed higher reading errors [11]. These results suggest that reading disability, a neurodevelopmental disorder, might be prevalent in boys with high TCDD and 1,2,3,4,6,7,8-HpCDD levels. In addition, the high 1,2,3,4,7,8-HxCDD group and high HpCDD group had lower mathematics achievement test scores, suggesting these boys may have mathematics learning impairment as well as language learning deficits.

Among girls 8 years of age, the high TCDD (≥3.0 pg/g lipid) and high TEQ-PCDD/Fs (≥17.6 pg TEQ/g lipid) groups displayed higher hyperactivity scores, suggesting increased ADHD traits [10]. However, no association was found between learning ability, assessed with the CLDQ, and the reading test in girls.

Thao et al. (2020) [12] investigated sexual dimorphism in gaze behavior in 8-year-old children and reported that feminine index scores, defined as longer fixation duration on girl-oriented pictures, compared to boy-oriented pictures, were significantly higher in boys with high TEQ-PCDD/Fs (≥17.6 pg TEQ/g lipid), for almost all PCDD congeners. In girls, however, the feminine index scores were significantly higher only in the high TCDD group (≥3.5 pg/g lipid).

Taken together, TCDD exposure may be associated with the occurrence of neurodevelopmental disorders, such as ASD, ADHD, and learning difficulties (LD). In boys, however, not only TCDD exposure but also PCDD congeners, such as 1,2,3,4,6,7,8-HpCDD and 1,2,3,4,7,8-HxCDD, may influence cognitive ability and reading ability and change their behavior. However, in girls, only high TCDD (≥3.5 pg/g lipid) may contribute to the occurrence of adverse effects on their brain functions and behavior.

Taken together, the findings suggest that TCDD exposure may be associated with neurodevelopmental disorders, such as ASD, ADHD, and learning difficulties. In boys, however, not only exposure to TCDD, but also exposure to PCDD congeners, such as 1,2,3,4,6,7,8-HpCDD and 1,2,3,4,7,8-HxCDD, may affect cognitive and reading abilities and modify behavior. In comparison, in girls, only those with high TCDD (≥3.5 pg/g lipid) may exhibit changes in brain functions and behavior.

4. Discussion

4.1. Neurodevelopmental Disorders and Perinatal TCDD Exposure

We followed up perinatally dioxin-exposed infants from the Da Nang birth cohort for 8 years and found that exposure to TEQ-PCDD/Fs influenced language and motor development, particularly in boys, in the first 3 years of life. At 5 years of age, poor motor skills were also observed in boys with high TEQ-PCDD/Fs. Notably, TCDD exposure was specifically associated with an increased occurrence of neurodevelopmental disorders without intellectual disability, such as ASD, ADHD, or specific learning disorders (LDs). However, 1,2,3,4,7,8-HxCDD and 1,2,3,4,6,7,8-HpCDD, which are TEQ-PCDD/F constituents, were also associated with reading LDs in boys.

4.1.1. ASD and Perinatal TCDD Exposure

ASD is a neurodevelopmental disorder that can be diagnosed around 2 years of age. However, previous studies in children exposed to high levels of dioxins and PCBs in the

Netherlands [2] and Taiwan [23,24] have reported associations of dioxin exposure with general neurodevelopment, but not neurodevelopmental disorders such as ASD. Because neurodevelopmental disorder in children was only categorized as a psychiatric disorder approximately 20 years ago, studies published before 2000 did not examine associations with disorders such as ASD in children. Furthermore, children with increased ASD risk cannot be screened using general neurodevelopmental test batteries such as Bayley III, which were developed for early detection in children with an intellectual disability.

In the Da Nang cohort survey, we used ASRS to evaluate ASD behavior based on criteria in the Diagnostic and Statistical Manual for Mental Disorders, 4th Edition, text revision (DSM-IV-TR), published in 2009. This scale has three subscales, including social communication and unusual behavior. The symptoms and T-scores for each scale are calculated from raw values after percentile rank conversion, and it is a sensitive tool for screening children with autistic traits.

We found that children with TCDD levels ≥3.5 pg/g lipid showed significantly higher scores in the ASRS scales at 3 years of age compared to children with low exposure, in both sexes [4]. However, at 5 years of age, unusual behavior scores were significantly higher only in girls with TCDD ≥ 3.0 pg/g lipid [10]. A reason why boys showed no significant association with TCDD at 5 years of age might be the smaller number of children with high TCDD (≥3.5 pg/g lipid) in the survey compared to 3-year-olds. Another reason might be changes in behavior in children with high TCDD because of increased ADHD symptoms, particularly hyperactivity, in boys at 5 years of age with high TCDD ≥ 3.0 (pg/g lipid) [10]. This speculation is consistent with the significantly lower cognitive ability, assessed with the NVI in KABC-II, in boys with high TCDD (≥3.5 pg/g lipid), shown in Table 3, as well as high TCDD (≥2.5 pg/g lipid) [9]. In contrast, girls at 5 years of age did not show increased ADHD symptoms or lower NVI scores associated with high TCDD exposure.

Table 3. Differences in toxin levels between sexes in children with high TCDD for each TEQ-PCDD/F category.

	TEQ-PCDD/Fs														
	<17.6					>17.6					All				
	Boys *N* = 104		Girls *N* = 73		*p*	Boys *N* = 24		Girls *N* = 26		*p*	Boys *N* = 128		Girls *N* = 99		*p*
Mean, SD of TEQ	10.6	1.4	10.5	1.5	0.826	22.5	1.2	23.3	1.3	0.620	12.2	1.6	12.9	1.7	0.392
Mean, SD of TCDD	1.2	2.0	1.1	2.3	0.749	2.6	1.7	3.2	1.9	0.236	1.3	2.1	1.5	2.5	0.430
N, (%) of TCDD > 2.5	12	(11.5)	8	(11.0)	0.904	10	(41.7)	18	(69.2)	0.050	22	(17.2)	26	(26.3)	0.098
N, (%) of TCDD > 3.0	5	(4.8)	4	(5.5)	0.842	7	(29.2)	11	(42.3)	0.332	12	(9.4)	15	(15.2)	0.184
N, (%) of TCDD > 3.5	3	(2.9)	3	(4.1)	0.660	5	(20.8)	10	(38.5)	0.171	8	(6.3)	13	(13.1)	0.086
N, (%) of TCDD > 5.5	0	(0.0)	0	(0.0)	-	2	(8.3)	7	(26.9)	0.079	2	(1.6)	7	(7.1)	0.035

Units: pg/g lipid for TCDD and pg-TEQ/g lipid for TEQ-PCDD/Fs, *N*: number of subjects, SD: standard deviation; TEQ: TEQ-PCDD/Fs, *p*: *p*-values compared between boys and girls by likelihood ratio test.

4.1.2. ADHD and Perinatal TCDD Exposure

ADHD is another neurodevelopmental disorder, and children with ASD are often diagnosed with ADHD as a comorbidity of ASD [25], and associations with exposure to endocrine disrupter chemicals, including PCB [26–29], pesticides [30,31], and dioxins [19,32], have been investigated in school-aged children. We also investigated associations between perinatal dioxin exposure and ADHD behaviors in children from the Da Nang cohort when they reached 8 years of age. Increased ADHD symptoms, particularly hyperactivity behaviors, were found in girls with TCDD ≥ 3.0 pg/g lipid or TEQ-PCDD/Fs ≥ 17.6 pg TEQ/g lipid, although they did not show increased ADHD symptoms at 5 years of age [10].

However, no increased aggressive behavior or ADHD associated with dioxin exposure was found in boys at 8 years of age, whose hyperactivity scores in the ADHD-RS were associated with TCDD exposure at 5 years of age.

Because it has often been observed in children with ADHD, aggressive behavior was also examined using the Children's Scale of Hostility and Aggression: Reactive/Proactive (C-SHARP) with five subscales (verbal aggression, bullying, covert aggression, hostility, and physical aggression). The prevalence of high covert aggression scores in children, particularly in girls, was significantly higher in the high TCDD group (≥3.0 pg/g lipid) [33], suggesting behavior problems related to ADHD in girls exposed to high levels of TCDD.

4.1.3. Learning Disorders (LDs) and Perinatal TCDD Exposure

Specific LD is a neurodevelopmental disorder that often co-occurs with other behavioral disorders in school-aged children [34,35]. CLDQ, which has two subscales—a math score and a reading score—is a good indicator of learning disability in school children.

An increased prevalence of LDs, based on interviewing parents, has been reported in school children exposed to background levels of dioxins and is associated with elevated serum levels of PCDDs and PCDF congeners [36]. Pham-The et al. (2020) [11] investigated the effects of perinatal dioxin exposure on learning difficulty using CLDQ and mathematics and language (Vietnamese) achievement tests in children from the Da Nang cohort when they reached school age. Boys with high TCDD (>3.5 pg/g lipid) or high 1,2,3,4,6,7,8-HpCDD (≥10.0 pg/g lipid) showed increased CLDQ reading scores, indicating a reading disability, such as dyslexia. Boys with high TCDD also showed significantly lower language achievement scores and higher errors and lower speed in reading a passage. High HpCDD and 1,2,3,4,7,8-HxCDD influenced not only reading learning but also mathematics learning, indicated by lower scores in mathematics achievement tests. These results suggest that TCDD exposure may mainly impair reading learning processes, including language processing, working memory, and processing speed control, which are collectively termed "lexical strategy". In contrast, PCDD/F congeners other than TCDD, such as 1,2,3,4,7,8-HxCDD and 1,2,3,4,6,7,8-HpCDD, may affect the learning process of both mathematics and language in boys, suggesting different brain areas may be affected by these compounds.

4.1.4. Gaze Behavior and Perinatal TCDD Exposure

Previous follow-up studies of children from birth cohorts in Duisburg, Germany [18] investigated the effects of total dioxin exposure on sex-typed play behavior in preschool children. These studies found increased feminine play behavior in boys and decreased masculine play behavior in girls, suggesting a feminization effect of dioxins, as endocrine disrupter chemicals. Thao et al. (2020) [12] examined gaze behavior in children at 8 years of age when viewing human line drawings, and found that feminine gaze behavior (preference for girl-oriented pictures) was increased in boys with high TEQ-PCDD/Fs (≥17.6 pg-TEQ/g lipid) and in girls with high TCDD (≥3.5 pg/g lipid). These investigators also examined salivary testosterone levels, an indicator of pubertal stage, in these children; however, an association of testosterone level with feminine gaze behavior was not found in either sex.

In boys, increased feminine gaze behavior, correlated with TEQ-PCDD/Fs, was found to be inversely correlated with hand movement scores and sequential index scores in the KABC-II examined at 5 years of age [9]. These findings suggest that impaired cognitive function, caused by dioxins perturbing brain development during the perinatal period, may lead to increased feminine gaze behavior in boys. In this study, girls with high TCDD (≥3.5 pg/g lipid) preferred female-oriented pictures, regardless of whether they were human or non-human, indicating an increased duration of fixed gaze behavior towards preferred pictures. This is often observed in children with ASD without an intellectual disability [37]. Moreover, in girls, lower cognitive scores related to facial recognition in the KABC-II examined at 5 years of age were associated with increased feminine gaze behavior observed at 8 years of age, suggesting poor facial recognition ability, which is often observed in children with ASD [38,39] in girls with high TCDD. These findings suggest that in girls, TCDD exposure may affect perinatal brain development and impair cognitive ability associated with social–emotional behavior, distinct from the cognitive functions affected in boys with high levels of TEQ-PCDD/Fs. Further studies are needed on the sex-specific effects of dioxins on social–emotional behavior during the adolescent period, when behaviors change drastically in both boys and girls.

4.2. Neurological Effects of Perinatal TCDD Exposure Detected by Electroencephalography (EEG)

In children 9 years of age in the Da Nang birth cohort, we investigated the effects of perinatal dioxin exposure on mu and theta rhythms by analyzing EEG recordings during hand movements, which reflect the activity of the mirror neuron system in the brain [40]. The mirror neuron system is reported to be impaired in children with ASD, resulting in poor social cognition. This concept is commonly called the "broken mirror theory" of autism [41]. In the studies in young (2–8-year-old) children, EEG power reduction in the theta band (4–8 Hz) during action observation or execution is a good indicator of mirror neuron activity [42]. In the Da Nang cohort, reduction in EEG power in the theta band caused by mirror neuron activity was significantly less in girls with high TCDD (≥3 pg/g lipid) and in boys with high TEQ-PCDD/Fs (≥17.6 pg TEQ/g lipid), particularly high HxCDDs and several PCDF congeners [40]. These results suggest that TCDD may be a congener that specifically impairs the mirror neuron system in the brain, resulting in social behavior problems, such as ADHD, in girls at 8 years of age [10]. However, in boys, not only TCDD, but also other PCDD/F congeners, may impact the neural substrates of cognitive functions, including the mirror neuron system, resulting in learning difficulties during school age.

Vu et al. (2021b) [43] performed magnetic resonance imaging (MRI) analysis in 32 men living in the most dioxin-contaminated area, originating from Agent Orange use near Bien Hoa Air Base in Vietnam, to investigate associations between dioxin exposure and brain structural irregularities. The volume of the left inferior frontal gyrus pars orbitalis, which participates in cognitive and social–emotional functions, was significantly lower in men exposed to Agent Orange, mainly TCDD, during the perinatal period. This suggests that TCDD may affect brain regions, leading to social cognitive deficits in men. In future studies, MRI imaging analysis should be performed in women in Bien Hoa to clarify the effects of TCDD on brain regions and on connectivity among different brain areas.

4.3. Estimated Threshold Values of TCDD for Significant Neurodevelopmental Problems

In the studies on children from the Da Nang cohort, the dioxin levels in breast milk were used as perinatal exposure markers, and cut-off values were calculated from the GM and GSD values in the breast milk of nursing mothers in unsprayed areas—3.5 pg/g lipid for the high TCDD group (GM × GSD^3) and 17.6 (pg TEQ/g lipid) for the high TEQ-PCDD/Fs group (GM × GSD^4) [5]. The effects of TCDD on KABC-II scores at 5 years of age and on ADHD-RS scores at 8 years of age were observed in children with TCDD ≥ 2.5 pg/g lipid [9] and TCDD ≥ 3.0 pg/g lipid [10], respectively, whereas the effects of TEQ-PCDD/Fs were only detected in children with TEQ-PCDD/Fs ≥ 17.6 pg TEQ/g lipid. These results suggest that the threshold value for TEQ-PCDD/Fs is approximately 18 pg TEQ/g lipid. In comparison, the estimation of TCDD threshold values is difficult.

In girls, neurotoxic effects appeared after 5 years of age and were specific to TCDD exposure [10]. Furthermore, the proportion of children with high TCDD levels was higher among girls compared to boys at different cut-off values, including 2.5–5.5 pg/g lipid (Table 3). Particularly, when the cut-off value for the high TCDD group was set at 5.5 pg/g lipid, the proportion of all children with high TCDD was significantly greater in girls compared to boys (p = 0.035). These findings suggest that more girls were exposed to extremely high levels of TCDD during the fetal period compared to boys, resulting in a higher frequency of TCDD-specific neurotoxic effects in girls. Interestingly, a lower sex ratio at birth (lower percentage of boys), associated with serum TCDD levels, was reported in residents exposed to high levels of TCDD following a chemical factory explosion in Seveso, Italy [44], suggesting that female fetuses are relatively resistant to TCDD toxicity compared to male fetuses.

Taken together, the findings suggest that girls exposed to high levels of TCDD may be able to grow up without neurodevelopmental problems, except social–emotional cognitive deficits, similar to children with ASD and ADHD without intellectual deficits. However, the number of children with high TCDD exposure is too small to estimate sex-specific threshold values for the neurotoxic effects of TCDD. Future studies on children from the Bien Hoa cohort, including more children exposed to high TCDD levels (≥5.5 pg/g lipid), may help address this shortcoming [45].

4.4. TCDD-Induced Neurotoxicity via the Aryl Hydrocarbon Receptor (AhR)

Most, if not all, of the toxic and biological effects of TCDD are mediated through the aryl hydrocarbon receptor (AhR), as are the effects of PCDD/F congeners [21]. Thus, the total TEQ-PCDD/F value is frequently used to estimate the total TCDD-like toxicity of all PCDD/Fs. Notably, however, our findings suggest that only TCDD increased the risk for ASD in 3-year-old children [4], LD in boys [11], and ADHD in girls at 8 years of age [10] after adjusting for confounding factors, such as maternal age, education, smoking of family members, drinking habit during pregnancy, and economic status.

As perinatal dioxin exposure markers, dioxins in maternal breast milk were used in the Da Nang birth cohort studies, of which relevant factors to each congener were investigated by Tai et al. (2011) [1] and Anh et al. (2014) [46]. The length of residency nearby the air base was the most important factor in increased dioxin levels in breast milk, although some food consumption increased several PCDD/Fs congeners. Particularly, TCDD concentrations were highly associated with only increased residency around Da Nang Air Base, suggesting that TCDD exposure may have originated from contamination of Agent Orange in the soil and sediment around Da Nang Air Base.

Marazziti et al. (2012) [47] reviewed the studies to investigate associations between mitochondrial alterations and neuropsychiatric diseases, including neurodevelopmental disorders, and suggested that mitochondrial abnormalities may have a role in the onset or pathophysiology in developmental disorders, such as autism and ADHD. Additionally, in a recent clinical study, Lee et al. (2019) [48] reported that only girls with ADHD showed higher levels of HtrA2 in plasma, a mitochondria-associated protein, compared to controls, and that their HtrA2 levels were inversely correlated with behavioral symptoms. These findings indicate that mitochondrial pathways may have an important role in the pathophysiology of ADHD in girls.

In an in vitro study, Hwang et al. (2016) [49] reported that a portion of the cellular pool of AhR was localized to the inter-membrane space of the mitochondria and that TCDD exposure induced degradation of the AhR pool in mitochondria, resulting in altered cellular respiration and influencing a battery of proteins associated with various metabolic pathways within the mitochondrial proteome. These results suggest that TCDD exposure may induce mitochondrial dysfunction via AhR, leading to cell energy metabolic abnormality.

Taken together, and compared to other PCDD congeners, TCDD might have more specific AhR-dependent toxicity to induce mitochondrial dysfunction, which is involved in the pathophysiology of neurodevelopmental disorders, such as ASD and ADHD, observed in children from the Da Nang cohort. Further studies are required to clarify the differences in neurotoxic effects between TCDD and other PCDD/Fs congeners and their mechanisms in the future.

5. Conclusions

Perinatal TCDD exposure affects social–emotional cognitive functions, leading to neurodevelopmental disorders. However, the effects are sex-specific: primarily LD in boys and predominantly ADHD in girls. PCDD/F congeners other than TCDD may specifically impact neurodevelopment in boys. The estimated threshold values for TCDD may also differ between the sexes. Further studies on children, including those with high TCDD exposure, are necessary to clarify the sex and age-specific neurodevelopmental effects of these dioxins.

Author Contributions: Conceptualization, N.N.T. and M.N.; methodology, N.N.T., T.N.P. and T.P.-T.; writing—original draft preparation, N.N.T., M.N. and H.T.V.; writing—review and editing, M.N. and T.N.P.; supervision, K.N.L. All authors have read and agreed to the published version of the manuscript.

Funding: This research received no external funding.

Institutional Review Board Statement: Not applicable.

Informed Consent Statement: Not applicable.

Data Availability Statement: Not applicable.

Acknowledgments: We thank all subjects participating in our studies. We are grateful to the medical staff at the Health Department of Da Nang city and the commune health centers of Thanh Khe and Son Tra districts for their collaboration in the surveys. We thank Barry Patel for editing a draft of this manuscript. We thank Barry Patel for editing a draft of this manuscript.

Conflicts of Interest: The authors declare no conflict of interest.

References

1. Tai, P.T.; Nishijo, M.; Kido, T.; Nakagawa, H.; Maruzeni, S.; Naganuma, R.; Anh, N.T.; Morikawa, Y.; Luong, H.V.; Anh, T.H.; et al. Dioxin concentrations in breast milk of Vietnamese nursing mothers: A survey four decades after the herbicide spraying. *Environ. Sci. Technol.* **2011**, *45*, 6625–6632. [CrossRef] [PubMed]
2. Koopman-Esseboom, C.; Weisglas-Kuperus, N.; de Ridder, M.A.; Van der Paauw, C.G.; Tuinstra, L.G.; Sauer, P.J. Effects of polychlorinated biphenyl/dioxin exposure and feeding type on infants' mental and psychomotor development. *Pediatrics* **1996**, *97*, 700–706. [CrossRef] [PubMed]
3. Nakajima, S.; Saijo, Y.; Kato, S.; Sasaki, S.; Uno, A.; Kanagami, N.; Hirakawa, H.; Hori, T.; Tobiishi, K.; Todaka, T.; et al. Effects of prenatal exposure to polychlorinated biphenyls and dioxins on mental and motor development in Japanese children at 6 months of age. *Environ. Health Perspect.* **2006**, *114*, 773–778. [CrossRef] [PubMed]
4. Nishijo, M.; Pham, T.T.; Nguyen, A.T.; Tran, N.N.; Nakagawa, H.; Hoang, L.V.; Tran, A.H.; Morikawa, Y.; Ho, M.D.; Kido, T.; et al. 2,3,7,8-Tetrachlorodibenzo-p-dioxin in breast milk increases autistic traits of 3-year-old children in Vietnam. *Mol. Psychiatry* **2014**, *19*, 1220–1226. [CrossRef]
5. Tai, P.T.; Nishijo, M.; Anh, N.T.; Maruzeni, S.; Nakagawa, H.; Van Luong, H.; Anh, T.H.; Honda, R.; Kido, T.; Nishijo, H. Dioxin exposure in breast milk and infant neurodevelopment in Vietnam. *Occup. Environ. Med.* **2013**, *70*, 656–662. [CrossRef]
6. Nishijo, M.; Tai, P.T.; Nakagawa, H.; Maruzeni, S.; Anh, N.T.; Luong, H.V.; Anh, T.H.; Honda, R.; Morikawa, Y.; Kido, T.; et al. Impact of perinatal dioxin exposure on infant growth: A cross-sectional and longitudinal studies in dioxin-contaminated areas in Vietnam. *PLoS ONE* **2012**, *7*, e40273. [CrossRef]
7. Pham, T.T.; Nishijo, M.; Nguyen, A.T.N.; Tran, N.N.; Hoang, L.V.; Tran, A.H.; Nguyen, T.V.; Nishijo, H. Perinatal dioxin exposure and the neurodevelopment of Vietnamese toddlers at 1 year of age. *Sci. Total Environ.* **2015**, *536*, 575–581. [CrossRef]
8. Tai, P.T.; Nishijo, M.; Nghi, T.N.; Nakagawa, H.; Van Luong, H.; Anh, T.H.; Nishijo, H. Effects of Perinatal Dioxin Exposure on Development of Children during the First 3 Years of Life. *J. Pediatr.* **2016**, *175*, 159–166.e2. [CrossRef]
9. Tran, N.N.; Pham, T.T.; Ozawa, K.; Nishijo, M.; Nguyen, A.T.; Tran, T.Q.; Hoang, L.V.; Tran, A.H.; Phan, V.H.; Nakai, A.; et al. Impacts of Perinatal Dioxin Exposure on Motor Coordination and Higher Cognitive Development in Vietnamese Preschool Children: A Five-Year Follow-Up. *PLoS ONE* **2016**, *11*, e0147655. [CrossRef]
10. Pham-The, T.; Nishijo, M.; Pham, T.N.; Vu, H.T.; Tran, N.N.; Tran, A.H.; Hoang, L.V.; Do, Q.; Nishino, Y.; Nishijo, H. Perinatal Dioxin Exposure and Attention Deficit Hyperactivity Disorder (ADHD) Symptoms in Children Living in a Dioxin Contamination Hotspot in Vietnam. *Toxics* **2022**, *10*, 212. [CrossRef]
11. Pham The, T.; Pham Ngoc, T.; Hoang Van, T.; Nishijo, M.; Tran Ngoc, N.; Vu Thi, H.; Hoang Van, L.; Tran Hai, A.; Nishino, Y.; Nishijo, H. Effects of perinatal dioxin exposure on learning abilities of 8-year-old children in Vietnam. *Int. J. Hyg. Environ. Health* **2020**, *223*, 132–141. [CrossRef] [PubMed]
12. Thao, P.N.; Nishijo, M.; Tai, P.T.; Hoa, V.T.; Nghi, T.N.; Anh, T.H.; Quyet, D.; Takiguchi, T.; Nishino, Y.; Nishijo, H. Dioxin exposure and sexual dimorphism of gaze behavior in prepubertal Vietnamese children living in Da Nang, a hot spot for dioxin contamination. *Sci. Total Environ.* **2020**, *749*, 141083.
13. Yim, G.; Minatoya, M.; Kioumourtzoglou, M.A.; Bellavia, A.; Weisskopf, M.; Ikeda-Araki, A.; Miyashita, C.; Kishi, R. The associations of prenatal exposure to dioxins and polychlorinated biphenyls with neurodevelopment at 6 Months of age: Multi-pollutant approaches. *Environ Res.* **2022**, *209*, 112757. [CrossRef] [PubMed]

14. Nakajima, S.; Saijo, Y.; Miyashita, C.; Ikeno, T.; Sasaki, S.; Kajiwara, J.; Kishi, R. Sex-specific differences in effect of prenatal exposure to dioxin-like compounds on neurodevelopment in Japanese children: Sapporo cohort study. *Environ Res.* **2017**, *159*, 222–231. [CrossRef]
15. Ikeno, T.; Miyashita, C.; Nakajima, S.; Kobayashi, S.; Yamazaki, K.; Saijo, Y.; Kita, T.; Sasaki, S.; Konishi, K.; Kajiwara, J.; et al. Effects of low-level prenatal exposure to dioxins on cognitive development in Japanese children at 42 months. *Sci. Total Environ.* **2018**, *618*, 1423–1430. [CrossRef]
16. Yamazaki, K.; Itoh, S.; Ikeda-Araki, A.; Miyashita, C.; Hori, T.; Hachiya, N.; Kishi, R. Association of prenatal exposure to dioxin-like compounds, polychlorinated biphenyl, and methylmercury with event-related brain potentials in school-aged children: The Hokkaido study. *Neurotoxicology* **2022**, *1*, 11–21. [CrossRef]
17. Winneke, G.; Ranft, U.; Wittsiepe, J.; Kasper-Sonnenberg, M.; Fürst, P.; Krämer, U.; Seitner, G.; Wilhelm, M. Behavioral sexual dimorphism in school-age children and early developmental exposure to dioxins and PCBs: A follow-up study of the Duisburg Cohort. *Environ. Health Perspect.* **2014**, *122*, 292–298. [CrossRef]
18. Wilhelm, M.; Wittsiepe, J.; Lemm, F.; Ranft, U.; Krämer, U.; Fürst, P.; Röseler, S.C.; Greshake, M.; Imöhl, M.; Eberwein, G.; et al. The Duisburg birth cohort study: Influence of the prenatal exposure to PCDD/Fs and dioxin-like PCBs on thyroid hormone status in newborns and neurodevelopment of infants until the age of 24 months. *Mutat. Res.* **2008**, *659*, 83–92. [CrossRef]
19. Neugebauer, J.; Wittsiepe, J.; Kasper-Sonnenberg, M.; Schöneck, N.; Schölmerich, A.; Wilhelm, M. The influence of low level pre- and perinatal exposure to PCDD/Fs, PCBs, and lead on attention performance and attention-related behavior among German school-aged children: Results from the Duisburg Birth Cohort Study. *Int. J. Hyg. Environ. Health.* **2015**, *218*, 153–162. [CrossRef]
20. Nowack, N.; Wittsiepe, J.; Kasper-Sonnenberg, M.; Wilhelm, M.; Schölmerich, A. Influence of Low-Level Prenatal Exposure to PCDD/Fs and PCBs on Empathizing, Systemizing and Autistic Traits: Results from the Duisburg Birth Cohort Study. *PLoS ONE.* **2015**, *10*, e0129906. [CrossRef]
21. Van den Berg, M.; Birnbaum, L.S.; Denison, M.; De Vito, M.; Farland, W.; Feeley, M.; Fiedler, H.; Hakansson, H.; Hanberg, A.; Haws, L.; et al. The 2005 World Health Organization Reevaluation of Human and Mammalian Toxic Equivalency Factors for Dioxins and Dioxin-Like Compounds. *Toxicol. Sci.* **2006**, *93*, 223–241. [CrossRef] [PubMed]
22. Patrick, K.E.; McCurdy, M.D.; Chute, D.L.; Mahone, E.M.; Zabel, T.A.; Jacobson, L.A. Clinical utility of the Colorado Learning Difficulties Questionnaire. *Pediatrics* **2013**, *132*, e1257–e1264. [CrossRef] [PubMed]
23. Jacobson, J.L.; Jacobson, S.W. Intellectual impairment in children exposed to polychlorinated biphenyls in utero. *N. Engl. J. Med.* **1996**, *335*, 783–789. [CrossRef] [PubMed]
24. Guo, Y.L.; Lambert, G.H.; Hsu, C.C.; Hsu, M.M. Yucheng: Health effects of prenatal exposure to polychlorinated biphenyls and dibenzofurans. *Int. Arch. Occup. Environ. Health* **2004**, *77*, 153–158. [CrossRef]
25. Lord, C.; Elsabbagh, M.; Baird, G.; Veenstra-Vanderweele, J. Autism spectrum disorder. *Lancet* **2018**, *392*, 508–520. [CrossRef] [PubMed]
26. Stewart, P.; Fitzgerald, S.; Reihman, J.; Gump, B.; Lonky, E.; Darvill, T.; Pagano, J.; Hauser, P. Prenatal PCB exposure, the corpus callosum, and response inhibition. *Environ. Health Perspect.* **2003**, *111*, 1670–1677. [CrossRef]
27. Stewart, P.; Reihman, J.; Gump, B.; Lonky, E.; Darvill, T.; Pagano, J. Response inhibition at 8 and 9 1/2 years of age in children prenatally exposed to PCBs. *Neurotoxicol. Teratol.* **2005**, *27*, 771–780. [CrossRef]
28. Sagiv, S.K.; Thurston, S.W.; Bellinger, D.C.; Tolbert, P.E.; Altshul, L.M.; Korrick, S.A. Prenatal organochlorine exposure and behaviors associated with attention deficit hyperactivity disorder in school-aged children. *Am. J. Epidemiol.* **2010**, *171*, 593–601. [CrossRef]
29. Sagiv, S.K.; Thurston, S.W.; Bellinger, D.C.; Altshul, L.M.; Korrick, S.A. Neuropsychological measures of attention and impulse control among 8-year-old children exposed prenatally to organochlorines. *Environ. Health Perspect.* **2012**, *120*, 904–909. [CrossRef]
30. Bouchard, M.F.; Bellinger, D.C.; Wright, R.O.; Weisskopf, M.G. Attention-deficit/hyperactivity disorder and urinary metabolites of organophosphate pesticides. *Pediatrics* **2010**, *125*, e1270–e1277. [CrossRef]
31. Marks, A.R.; Harley, K.; Bradman, A.; Kogut, K.; Barr, D.B.; Johnson, C.; Calderon, N.; Eskenazi, B. Organophosphate pesticide exposure and attention in young Mexican-American children: The CHAMACOS study. *Environ. Health Perspect.* **2010**, *118*, 1768–1774. [CrossRef] [PubMed]
32. Ames, J.; Warner, M.; Siracusa, C.; Signorini, S.; Brambilla, P.; Mocarelli, P.; Eskenazi, B. Prenatal dioxin exposure and neuropsychological functioning in the Seveso Second Generation Health Study. *Int. J. Hyg. Environ. Health* **2019**, *222*, 425–433. [CrossRef] [PubMed]
33. Pham-The, T.; Nishijo, M.; Pham-Ngoc, T.; Vu-Thi, H.; Tran-Ngoc, N.; Tran-Hai, A.; Hoang-Van, L.; Nishino, Y.; Nishijo, H. Effects of prenatal dioxin exposure on children behaviors at 8 years of age of age. In Proceedings of the 39th International Symposium on Halogenated Persistent Organic Pollutants—Dioxin, Kyoto, Japan, 25–30 August 2019; p. 517.
34. Curtis, P.R.; Frey, J.R.; Watson, C.D.; Hampton, L.H.; Roberts, M.Y. Language Disorders and Problem Behaviors: A Meta-analysis. *Pediatrics* **2018**, *142*, e20173551. [CrossRef] [PubMed]
35. Langberg, J.M.; Vaughn, A.J.; Brinkman, W.B.; Froehlich, T.; Epstein, J.N. Clinical utility of the Vanderbilt ADHD Rating Scale for ruling out comorbid learning disorders. *Pediatrics* **2010**, *126*, e1033–e1038. [CrossRef]
36. Lee, D.H.; Jacobs, D.R.; Porta, M. Association of serum concentrations of persistent organic pollutants with the prevalence of learning disability and attention deficit disorder. *J. Epidemiol. Community Health* **2007**, *61*, 591–596. [CrossRef]
37. Falck-Ytter, T.; Bolte, S.; Gredeback, G. Eye tracking in early autism research. *J. Neurodev. Disord.* **2013**, *5*, 28. [CrossRef]

38. Klin, A.; Sparrow, S.S.; de Bildt, A.; Cicchetti, D.V.; Cohen, D.J.; Volkmar, F.R. A normed study of face recognition in autism and related disorders. *J. Autism Dev. Disord.* **1999**, *29*, 499–508. [CrossRef]
39. Hedley, D.; Brewer, N.; Young, R. Face recognition performance of individuals with Asperger syndrome on the Cambridge Face Memory Test. *Autism Res.* **2011**, *4*, 449–455. [CrossRef]
40. Vu, H.T.; Nishijo, M.; Pham, T.N.; Pham-The, T.; Hoanh, L.V.; Tran, A.H.; Tran, N.N.; Nishino, Y.; Do, Q.; Nishijo, H. Effects of perinatal dioxin exposure on mirror neuron activity in 9-year-old children living in a hot spot of dioxin contamination in Vietnam. *Neuropsychologia* **2021**, *161*, 108001. [CrossRef]
41. Hamilton, A.F. Reflecting on the mirror neuron system in autism: A systematic review of current theories. *Dev. Cogn. Neurosci.* **2013**, *3*, 91–105. [CrossRef]
42. Cochin, S.; Barthelemy, C.; Roux, S.; Martineau, J. Electroencephalographic activity during perception of motion in childhood. *Eur. J. Neurosci.* **2001**, *13*, 1791–1796. [CrossRef] [PubMed]
43. Vu, H.T.; Pham, T.N.; Yokawa, T.; Nishijo, M.; The, T.P.; Do, Q.; Nishino, Y.; Nishijo, H. Alterations in Regional Brain Regional Volume Associated with Dioxin Exposure in Men Living in the Most Dioxin-Contaminated Area in Vietnam: Magnetic Resonance Imaging (MRI) Analysis Using Voxel-Based Morphometry (VBM). *Toxics* **2021**, *9*, 353. [CrossRef] [PubMed]
44. Mocarelli, P.; Gerthoux, P.M.; Ferrari, E.; Patterson, D.G., Jr.; Kieszak, S.M.; Brambilla, P.; Vincoli, N.; Signorini, S.; Tramacere, P.; Carreri, V.; et al. Paternal concentrations of dioxin and sex ratio of offspring. *Lancet* **2000**, *355*, 1858–1863. [CrossRef] [PubMed]
45. Pham, N.T.; Nishijo, M.; Pham, T.T.; Tran, N.N.; Le, V.Q.; Tran, H.A.; Phan, H.A.V.; Nishino, Y.; Nishijo, H. Perinatal dioxin exposure and neurodevelopment of 2-year-old Vietnamese children in the most contaminated area from Agent Orange in Vietnam. *Sci. Total Environ.* **2019**, *678*, 217–226. [CrossRef]
46. Anh, N.T.; Nishijo, M.; Tai, P.T.; Maruzeni, S.; Morikawa, Y.; Anh, T.H.; Van Luong, H.; Dam, P.M.; Nakagawa, H.; Son Le, K.; et al. Maternal risk factors associated with increased dioxin concentrations in breast milk in a hot spot of dioxin contamination in Vietnam. *J. Expo. Sci. Environ. Epidemiol.* **2014**, *24*, 489–496. [CrossRef]
47. Marazziti, D.; Baroni, S.; Picchetti, M.; Landi, P.; Silvestri, S.; Vatteroni, E.; Catena Dell'Osso, M. Psychiatric disorders and mitochondrial dysfunctions. *Eur. Rev. Med. Pharmacol. Sci.* **2012**, *16*, 270–275.
48. Lee, C.J.; Wu, C.C.; Chou, W.J.; Lee, M.J.; Chou, M.C.; Lee, S.Y.; Wang, L.J. Mitochondrial-associated protein biomarkers in patients with attention-deficit/hyperactivity disorder. *Mitochondrion* **2019**, *49*, 83–88. [CrossRef]
49. Hwang, H.J.; Dornbos, P.; Steidemann, M.; Dunivin, T.K.; Rizzo, M.; LaPres, J.J. Mitochondrial-targeted aryl hydrocarbon receptor and the impact of 2,3,7,8-tetrachlorodibenzo-p-dioxin on cellular respiration and the mitochondrial proteome. *Toxicol. Appl. Pharmacol.* **2016**, *304*, 121–132. [CrossRef]

toxics MDPI

Review

Effects of Polychlorinated Dibenzo-p-dioxins, Polychlorinated Dibenzofurans, and Dioxin-like PCBs on Teeth and Bones in Animals and Humans

Tomoya Takiguchi, Hoa Thi Vu * and Yoshikazu Nishino

Department of Public Health, Kanazawa Medical University, Uchinada 920-0293, Japan
* Correspondence: vuhoa5593hvqy@gmail.com; Tel.: +81-762-188-093

Abstract: Bone metabolism is regulated by endocrine systems, so people exposed to polychlorinated dibenzo-p-dioxins and dibenzofurans (PCDD/Fs) may suffer adverse effects on bones and teeth. We reviewed previous publications in which effects of PCDD/Fs and dioxin-like polychlorinated biphenyls on the teeth and bones of animals and humans were found. The aim was to identify future research directions, particularly for epidemiological studies of populations exposed to PCDD/Fs in the environment. Exposure of fetuses to PCDD/Fs may affect odontogenesis, particularly enamel formation, but the effects of PCDD/Fs on bone genesis are limited to palatine bone. Exposure to PCDD/Fs in milk may affect both teeth and bones, but the effects on bones may be reversible. Exposure to high PCDD/F concentrations even during adulthood may adversely affect teeth. Exposure to PCDD/Fs may induce osteogenesis and improve bone properties because the disrupting effects of PCDD/Fs cause bone remodeling and vitamin D activation. More studies involving humans are required to investigate previously found associations between the PCDD/F concentrations humans are exposed to and biological markers for teeth and bones, including metabolites of vitamin D.

Keywords: dioxins; dioxin-like PCB; animals; human; teeth; bone

Citation: Takiguchi, T.; Vu, H.T.; Nishino, Y. Effects of Polychlorinated Dibenzo-p-dioxins, Polychlorinated Dibenzofurans, and Dioxin-like PCBs on Teeth and Bones in Animals and Humans. *Toxics* **2023**, *11*, 7. https://doi.org/10.3390/toxics11010007

Academic Editor: Michael Petriello

Received: 15 November 2022
Accepted: 17 December 2022
Published: 21 December 2022

1. Introduction

There are 75 polychlorinated dibenzo-p-dioxin (PCDD) congeners and 135 polychlorinated dibenzofuran (PCDF) congeners. Together, PCDDs and PCDFs are often abbreviated as PCDD/Fs. Each PCDD congener has between one and eight chlorine atoms attached to a dibenzo-p-dioxin structure, and each PCDF congener has between one and eight chlorine atoms attached to a dibenzofuran structure. However, only congeners with chlorine atoms in the 2, 3, 7, and 8 positions are extremely potent/toxic and highly persistent to humans and most animals. There are seven PCDD congeners and 10 PCDF congeners with chlorine atoms at the 2, 3, 7, and 8 positions. These 17 PCDD/F congeners should therefore be determined in samples from humans. There are 209 polychlorinated biphenyl (PCB) congeners, but only mono-ortho-chlorinated biphenyls and non-ortho-chlorinated biphenyls have similar types of toxicities to PCDD/Fs. These PCBs are called dioxin-like (dl) PCBs. Non-ortho-chlorinated biphenyls are more toxic than mono-ortho-chlorinated biphenyls. Only four non-ortho-chlorinated biphenyl congeners have therefore been determined in some studies.

2,3,7,8-Tetrachlorodibenzo-p-dioxin (TCDD) is the most toxic PCDD/F congener and is a by-product of the production of herbicides, such as 2,4-dichloroacetophenol and 2,4,5-trichloroacetophenol. TCDD causes chloracne in occupationally exposed people. Suskind and Hertzberg (1984) performed a clinical epidemiological study aimed at identifying the health effects caused by industrial exposure of workers involved in manufacturing 2,4,5-trichloroacetophenol (which was contaminated with TCDD) [1]. They found a higher prevalence of chloracne (55.7%) associated with actinic elastosis of the skin in 204 exposed workers than in non-exposed workers [1].

A three-year-old girl was found to have chloracne in 1968, and many people in southern parts of Japan were subsequently found to have Yusho disease, which was caused by the consumption of rice oil contaminated with PCBs and PCDFs. Yamashita and Hayashi (1985) found that infants exposed to PCBs as fetuses had peculiar clinical symptoms including cola-colored skin, gingival hyperplasia, natal teeth, a large fontanelle, and unusual calcification of the skull bone (fetal PCB syndrome) [2]. In the same parts of Japan, lower body weights and heights were found for girls without Yusho symptoms but with mothers with breast milk containing high PCB concentrations compared to unexposed girls [2]. This indicated that organochlorine compounds, such as PCBs and PCDD/Fs, may alter calcium metabolism through endocrine disruption and cause a decrease in body size.

In the 1990s, the effects of PCDD/Fs on teeth in animals and humans were investigated in Finland, and developmental enamel defects were found to be the most sensitive effects of PCDD/Fs exposure ever known, and found to be important and useful markers of exposure to PCDD/Fs and dioxin-like compounds in humans [3,4]. However, very few epidemiological research works have investigated the effects of PCDD/Fs on teeth using markers of developing teeth [5,6].

Bone metabolism is regulated by endocrine systems, particularly estrogen signaling, so PCDD/Fs were expected to adversely affect bones in animals and humans. However, PCDD/Fs have not been found to adversely affect bones (e.g., by decreasing bone mass or increasing osteoporosis occurrence) in people exposed to high concentrations of PCDD/Fs and dl-PCBs, such as women accidentally exposed to high TCDD concentrations in Seveso, Italy [7].

In this review, we summarize previous reports of the effects of PCDD/Fs and dl-PCBs on teeth and bones in animals and humans and assess the differences between the results to allow future directions for studies of populations exposed to PCDD/Fs to be suggested.

2. Materials and Methods

Eligible studies were identified by searching the literature using Pub Med in English. The search terms were "dioxins" OR "TCDD" OR "dioxin like PCBs" AND "bone" OR "teeth". Only original articles were selected, and the studies described in the articles were divided into two groups after excluding in vitro studies: in vivo animal studies and human studies (including epidemiological and clinical studies) of teeth and bones. The animal and human studies were divided into two groups based on the exposure period: the first group was exposure during the fetal and lactational period, and the second group was exposure during adulthood including the juvenile period. This was because the sensitivity of animals and humans to toxicants' exposure is highly dependent on the stage of development.

3. Results

3.1. *Effects of PCDD/Fs and dl-PCBs on Tooth Growth and Oral Health*

3.1.1. Animal Studies

The effects of gestational and lactational exposure to TCDD on tooth growth and the effects of exposure to high TCDD concentrations in adulthood on tooth growth were investigated in some animal studies aimed at investigating the adverse health effects associated with exposure to PCDD/Fs (Table 1).

Effects of Gestational and Lactational Exposure to TCDD on Tooth Growth

Kattainen et al. (2001) investigated the effects of gestational and lactational exposure to low maternal TCDD doses (0.03–1.0 μg/kg) on tooth development (mainly third molars, which develop from the perinatal period to about six weeks old) in three strains of rats with different sensitivities to TCDD and different aryl hydrocarbon receptor (AhR) structures [8]. The third molar eruption inhibition rate caused by exposure to TCDD at the same dose was much higher in pups of the most susceptible strain than pups of the other strains [8]. TCDD at 1 mg/kg completely prevented the development of the third lower molars in 60% of males and 50% of females in the most sensitive rat line. The percentage of third molars that had erupted in five-week-old pups of the most susceptible strain decreased as the TCDD dose increased. The sizes of the erupted molars in pups of all of the strains (i.e., regardless of AhR sensitivity) decreased as the TCDD dose increased [8].

Table 1. Effects of TCDD on teeth in animals.

No.	Authors	Publication Year	Animals	Exposure Time	Exposure Dose of TCDD	Health Effects (Outcomes)
			Exposure during fetal period and/or lactation			
1	Kattainen et al. [8]	2001	Han/Wistar (H/W) and Long-Evans (L/E) rats	Gestation day (GD) 15	0.03–1 μg/kg of a single oral dose	Lower percentage of erupted third molars (60% in males and 50% in females exposed to TCDD at 1 mg/kg) and smaller molar size
2	Lukinmaa et al. [9]	2001	H/W rats	1 day after delivery	50 or 1000 μg/kg	Abnormal development of molars and arrested dentin formation in lower incisors
3	Miettinen et al. [10]	2002	Dioxin-sensitive rats	GD11, GD13, GD19, postnatal day (PND)0, PND2, PND4	1 μg/kg	Accelerated eruption of the lower incisors and retarded eruption of the third molars.
4	Gao et al. [11]	2004	H/W rats	The day after delivery	50 or 1000 μg/kg	Enamel matrix stagnation and thicker predentin in the first and second molars
			Exposure during adulthood			
1	Alaluusua et al. [12]	1993	TCDD-resistant H/W young adult male rats	11–12 weeks of age	1000 μg/kg of a single intra peritoneal dose	Thinner upper and lower incisors 16 weeks after exposure
2	Kiukkonen et al. [13]	2002	Resistant H/W female rats and susceptible L/E female rats	10–30 weeks of age	0. 17, 1.7, 17, 170 (H/W rats only) μg/kg	Mesenchymal and epithelial elements formation of incisors formation

H/W: Han/Wistar; L/E: Long/Evans; GD: Gestational day; PND: Postnatal day.

Lukinmaa et al. (2001) administered TCDD (at a dose of 50 or 1000 μg/kg) to PCDD/F-resistant Han/Wistar (H/W) rats on postnatal day 1 and investigated the radiographical and histological differences in tooth development in the pups on postnatal days 9 and 22 [9]. Third molars were more frequently missing and more molars in the maxilla were missing in the high-TCDD-dose (1000 μg/kg) group than the low-TCDD-dose group (50 μg/kg) [9]. Eruption of the third molars was delayed and the erupted third molars were not calcified (determined by radiographic examination) in both the high- and low-TCDD-dose groups [9]. First and second molar root formation ceased early but eruption was not affected. Dentin formation in the erupted mandibular incisors stopped at the pre-eruption stage because of pulp cell death [9]. These results suggest that the first, second, and third molars and the incisors are affected by lactational exposure to TCDD even in strains less sensitive to TCDD/AhR.

Miettinen et al. (2002) determined the critical window during gestational and lactational exposure to TCDD for effects on development of molars in rat pups using a maternal TCDD dose of 1 μg/kg [10]. The pups exposed to TCDD on days 11, 13, and 19 of gestation had missing third molars, and earlier lower incisor eruption was found at 2 days old, indicating that TCDD accelerated incisor eruption but delayed third molar eruption [10]. The most sensitive period for effects on third molar development was the early tooth formation stage (11–19 days of gestation) [10].

Gao et al. (2004) administered TCDD (at a dose of 50 or 1000 μg/kg) to H/W rats the day after birth and found enamel matrix stagnation (retention) and thicker predentin in the upper first and second molars of the exposed rats (at both doses) than of control rats [11]. Observations of immune-stained ameloblasts and odontoblasts suggested that TCDD inhibited tooth calcification by decreasing CYP1A1 activity via the AhR [11].

Thus, gestational and lactational exposure to TCDD affects the development of molars in rats, resulting in defects, delayed eruption, and impaired calcification. It has been found that the most sensitive period to TCDD exposure is the early stage of tooth morphogenesis in utero and that the mechanisms involved include ameloblast and odontoblast inhibition preventing calcification of the teeth via the AhR.

Effects of Chronic Exposure to TCDD at High Doses in Adulthood on Teeth

Alaluusua et al. (1993) examined the continually erupting portal teeth of young male rats 16 weeks after administering a single TCDD dose at 1000 μg/kg [12]. The upper and lower incisors were markedly thinner in the exposed group than the control group, and the pulp in the lower incisors in the exposed group was lingually exposed to the oral cavity at the incisor ends [12]. The labial surfaces of the incisors were grayish and mottled in the exposed group but brown in the control group [12]. Histological examination indicated that the pulp chambers in the affected incisors were larger (at the expense of dentin) in the exposed group than the control group [12]. The odontoblasts on the incisal sides gradually lost polarity and the pulp tissue became necrotic in the exposed group, and the dentin next to the pulp chamber was irregular [12]. The roots in the exposed group were markedly tapered and had mesiodistally flattened appearances. The superficial enamel layers in the exposed group were pigmented [12]. These results indicated that chronic exposure to TCDD affects all tissues of the incisors in young male rats.

Kiukkonen et al. (2002) exposed TCDD-resistant H/W rats and susceptible female Long–Evans (L/E) rats to TCDD at total doses of 0.17, 1.7, 17, and 170 μg/kg given as weekly doses for 20 weeks from 10 weeks of age and observed the effects on the incisors [13]. Exposure of both strains to TCDD doses of 17 and 170 μg/kg resulted in color defects and pulpal perforation in the lower incisors, and periodontal ligament and pulp cell necrosis and consequent arrest of dentin formation at the incisal ends were found during histological examination [13]. The incisors of H/W rats were more affected, resulting in larger perforations at TCDD doses ≥17 μg/kg than at TCDD doses < 17 μg/kg. The enamel-producing cells underwent early squamous epithelialization and marked proliferation accompanied by enamel discoloration [13]. These results suggested that relatively high TCDD doses affected both the mesenchymal and epithelial elements of the forming tooth. However, incisor tooth formation impairment was not markedly different for the rat strains with different AhR sensitivities [13].

Thus, exposure to TCDD at relatively high doses in adulthood affects both the mesenchymal and epithelial elements (the two elements involved in tooth formation) of the incisors, resulting in dentin formation ceasing and qualitative changes in the enamel. However, pathways other than AhR signaling may mediate the effects of TCDD on teeth in adult rats.

3.1.2. Studies of Humans: Epidemiological Studies

In some studies, natal teeth were found at birth in infants with fetal PCB syndrome (called Yusho disease) in Japan [2,14]. Infants with fetal PCB syndrome (called Yu-cheng disease) in Taiwan had similar symptoms to infants with Yusho disease in Japan, and natal teeth were found at birth [15,16]. However, no symptoms involving teeth were found in follow-up studies of children with Yu-cheng disease [17]. No greater number of natal and neonatal teeth were found in children exposed to PCBs than in non-exposed children in Finland [18]. Adverse effects on teeth other than natal teeth in patients with Yusho or Yu-cheng disease are therefore described below (Table 2).

Fetal and Lactational Exposure to PCDD/Fs

Alaluusua et al. (1996) investigated the association between exposure to PCDD/Fs in breast milk and enamel dysplasia in Finnish children aged 6–7 years and found that exposure to PCDD/Fs was associated with increased frequency and severity of enamel dysplasia in the permanent first molars mineralized during the first 2 years of life [4]. The association with enamel defects was stronger for PCDD/Fs than PCBs, suggesting that developing teeth might be good indicators of adverse health effects caused by exposure to PCDD/Fs early in life.

Table 2. Effects of dioxins and PCBs on teeth in humans.

No.	Author	Publication Year	Country/Region	Target Population (Gender, Age, Residents/Patients)	Exposure Indicators	Health Effects (Outcomes)
			Dioxin exposure during fetal and/or lactation period			
1	Alaluusua et al. [4]	1996	Finland	Children aged 6–7 years (residents)	PCDD/Fs in maternal breast milk	Enamel dysplasia of permanent first molars.
2	Laisi et al. [5]	2008	Finland	Mothers and their 167 children	PCDD/Fs and PCBs in placentas	No effects on the child's molars.
			Dioxin exposure during childhood and adulthood			
1	Fukuyama et al. [19]	1979	Japan	Patients with Yusho (oil disease in Japan) aged at 0 to 15 years	Patients diagnosed as Yusho disease	Delayed eruption of permanent teeth, abnormal number of teeth, and abnormal root shape.
2	Guo et al. [20]	1999	Taiwan	Patients with Yu-cheng (oil disease in Taiwan)	Patients and controls	High prevalence of broken teeth (reported clinical history) 14 years after of diagnosis.
3	Alaluusua et al. [6]	2004	Seveso, Italy	Residents exposed to TCDD from industrial explosion	Serum TCDD concentration soon after accident	Enamel dysplasia and defects of teeth.
4	Kanagawa et al. [21]	2008	Japan	Patients with Yusho (oil disease in Japan)	Serum levels of PCBs, PCDFs, and polyquarterphenyls (PCQs)	Tooth pigmentation 30 years after of diagnosis.
5	Ngoc et al. [22]	2019	Vietnam	Adults living in the herbicide-sprayed and non-sprayed areas	Residential areas (cases and controls)	Prevalence of enamel dysplasia.

Laisi et al. (2008) investigated associations between the occurrence and severity of hypo-mineralization of molars and incisors and exposure to PCDD/Fs and PCBs (indicated by the PCDD/F and PCB concentrations in the placentas) for Finnish mother–infant pairs, but no associations were found [5].

Few epidemiological studies have been performed to investigate associations between fetal and/or lactational exposure to PCDD/Fs and growing teeth in infants. The effects of exposure of children to PCDD/Fs and dl-PCBs using a birth cohort with quantified perinatal exposure should be studied.

Exposure to PCDD/Fs in Childhood and Adulthood

Fukuyama et al. (1979) investigated tooth development in patients with Yusho disease [19]. Tooth eruption was delayed in 18% of patients with oiliness from 0 to 14 years of age [19]. Delayed tooth eruption was a standard symptom in the diagnostic criteria for Yusho disease [21]. Abnormal root morphology was found in 77% of patients aged 0 to 15 years [19]. These results suggested that exposure to PCBs may affect tooth development in children.

Guo et al. (1999) monitored patients with Yu-cheng disease and controls for 14 years after the disease outbreak and compared the prevalence of reported problems, including gum pigmentation and broken teeth, diagnosed in a clinic or hospital [20]. Broken teeth were around two times as prevalent in patients of both sexes than the controls, and the odds ratios for gum pigmentation were much higher for both sexes (5.6 for men and 8.5 for women) [20]. However, it was not clear why teeth were easily broken.

Alaluusua et al. (2004) investigated dental and oral abnormalities in an area contaminated because of an explosion at a pesticide plant in Seveso, Italy, and in a lightly contaminated area [6]. Of subjects with enamel dysplasia, 93% were <5 years old at the time of the accident [6]. In this age group, 42% of the subjects with defects lived in contaminated areas and 26% lived in lightly contaminated areas, and the defects were associated with the TCDD concentrations in serum [6]. Tooth defects were found in 12.5% of the subjects living in polluted areas and 4.6% of the subjects living in lightly polluted areas and were associated with the TCDD concentrations in serum [6].

Kanagawa et al. (2008) monitored Yusho patients for >30 years and investigated associations between symptoms including tooth pigmentation and the PCB, PCDF, and polychlorinated quarterphenyl (PCQ) concentrations in serum by performing logistic regression analysis after selecting symptoms by performing principal component analysis [21]. High PCQ concentrations in serum ($\geq$0.10 μg/L) significantly increased the risk of tooth pigmentation, suggesting that exposure to PCQs may negatively affect teeth even a long time after exposure ceases [21].

Ngoc et al. (2019) performed a case–control study to investigate the effects of PCDD/Fs on the prevalence of enamel dysplasia in 2200 adults living in herbicide-sprayed and non-sprayed areas in Viet Nam [22]. The results indicated that enamel dysplasia occurred in 20.5% and 5.8% of adults living in the sprayed areas and the non-sprayed areas, respectively, suggesting that enamel dysplasia was almost twice as prevalent in the exposed population than in the control population [22]. Enamel dysplasia was more prevalent in the premolars than in the molars. Most lesions were found on the buccal surfaces of the teeth [22].

The results described above indicated that tooth loss, broken teeth, and enamel defects were more likely to occur in people exposed to high PCDD/F and PCB concentrations (e.g., Yusho patients and residents of Seveso, Italy) than in people not exposed to high PCDD/F and PCB concentrations. In the study performed in Viet Nam, only the prevalence of enamel dysplasia in exposed and unexposed populations was compared, and dose–response relationships between exposure to PCDD/Fs and tooth problems were not investigated. Studies of populations with quantified exposure to PCDD/Fs need to be performed to clarify the effects of exposure to PCDD/Fs on teeth and gums.

3.2. *Effects of PCDD/Fs and dl-PCBs on Bone Growth and Remodeling*

3.2.1. Animal Studies

Associations between exposure to high TCDD doses and bone development during the perinatal or lactational periods and during adulthood have been investigated. Dental development primarily occurs during pre- and early postnatal periods, whereas bone development continues through adolescence, with maximum bone mass occurring in the juvenile period and bone remodeling continuing throughout adulthood. Morphological and biochemical effects on bone with chronic exposure to TCDD were therefore investigated in all periods of life (Table 3).

Effects of Fetal and Lactational Exposure to TCDD on Bone Growth

Miettinen et al. (2005) investigated the effects of fetal and lactational exposure to TCDD on the bones of three strains of rats with different AhR susceptibilities [23]. In the most susceptible strain, exposure to TCDD at a maternal dose of 1 μg/kg decreased bone length, cortical cross-sectional area, and bone density but did not decrease flexural fracture strength and tibia, femur, and femoral neck stiffness [23]. These effects were more severe in pups' exposure earlier than later. No effects were found when only fetal exposure occurred, suggesting that lactational exposure was essential for bone impairment [23]. Most of the abnormalities were reversed in the first year of life, suggesting that most of the effects were reversible.

Nishimura et al. (2009) administered TCDD at a dose of 15 μg/kg orally to mouse dams on day 1 after delivery and found that the 1,25-dihydroxy vitamin D3 concentration doubled, and osteocalcin, collagen type 1, and alkaline phosphatase gene expression decreased markedly [24]. Abnormal calcification of the tibia, decreased osteoblastic osteogenic activity, and increased fibroblast growth factor 23 occurred in the exposed pups [24]. Through histomorphometry, it was found that TCDD altered osteoblastic activity but did not alter bone resorption activity [24]. The most prominent bone lesions were increased osteoid volume and thickness in the cortical and trabecular bones [24]. These results suggested that upregulation of vitamin D induced by exposure to TCDD decreases osteoblast activity, resulting in impaired bone mineralization.

Table 3. Effects of dioxins on bones in animals.

No.	Author	Publication Year	Experimental Animals	Exposure Time	Exposure Dose	Health Effects (Outcomes)
			Exposure during fetal period and/or lactation			
1	Miettinen et al. [23]	2005	Rat strains with different susceptibility	GD11, GD13, GD19, PND0, PND2, PND4	0.03, 0.1, 0.3, and 1 µg/kg of single oral doses to dams	Decreased bone length, cortical cross-sectional area, and bone density of tibia, femur, and femoral neck.
2	Nishimura et al. [24]	2009	C57BL/6J mice	day 1 after delivery	15 µg/kg of a single oral dose to dams	Abnormal calcification of the tibia, increased 1,25-dihydroxy vitamin D3, and decreased osteogenic biomarker activity. No alteration of bone resorptive marker activity.
3	Finnilä et al. [25]	2010	Female Sprague Dawley rats	GD 11	1 µg/kg of a single dose in gavage	Normal bone development, such as decreased plasticity, increased dynamic hardness, storage modulus, and composite modulus.
4	Yamada et al. [26]	2014	Pregnant ICR strain mice	Embryonic Day 12.5	40 µg/kg of a single dose in gavage	Palatal osteogenesis and myogenesis related with occurrence of cleft palates.
			Exposure during adulthood			
1	Alaluusua et al. [12]	1993	Young adult male H/W rats	11–12 weeks of age	1000 µg/kg of a single intraperitoneal dose	Smaller skull size
2	Jamsa et al. [27]	2001	H/W and L/E rats	10–30 weeks of age	1.7 to 170 µg/kg of percutaneous doses	Decreased tibia size, 3-point bending fracture force, and stiffness of the tibia, but no alteration of bone mineral density (BMD).
3	Herlin et al. [28]	2010	Female L/E and H/W rats	10–30 weeks of age	0, 0.14, 1.4, 14 and 140 µg/kg for total dose of percutaneous doses	Altered bone geometry and bone biomechanical parameters, but no effect on bone mineral density parameters.
4	Herlin et al. [29]	2013	AhR knockout (Ahr(−/−)) and wild-type (Ahr(+/+)) mice	18–22 weeks of age	200 µg/kg for total dose in gavage	Harder bone matrix, thinner cortical bone, mechanically weaker bone, and increased trabecular bone volume fraction.
5	Fader et al. [30]	2018	Male and female C57BL/6 mice	PND25-53 (every 4 days)	0.01–30 µg/kg of oral doses	Increased trabecular bone volume of femur bone, inhibition of bone resorption markers, increased number of osteoblasts on the trabecular bone surface, increased regulator of osteoblast differentiation and mineralization, increased serum 1,25-dihydroxy Vitamin D3.

GD: Gestation day; PND: Postnatal day; H/W: Han/Wistar; L/E: Long/Evans.

Finnila et al. (2010) studied the effects of perinatal exposure to TCDD on bone matrix maturation in rats [25]. Pregnant rats received an intragastrical dose of 1 µg/kg of TCDD on gestation day 11, and the tibia properties of the offspring were assessed on postnatal days 35 and 70 [25]. Normal development, such as decreased plasticity and increased dynamic hardness, storage modulus, and composite modulus, occurred in the controls, and changes related to normal development did not occur in the offspring exposed to TCDD, indicating that bone matrix maturation was delayed [25]. Decreased bone calcification, shortened tibia length, and decreased strength were observed in the exposed pups [25]. These results suggested that TCDD-induced decreases in bone strength are associated with changes in bone mineralization and shape rather than changes in the bone matrix.

Yamada et al. (2014) administered TCDD at a dose of 40 μg/kg to mouse dams at 12.5 d of gestation and investigated the effects of exposure to TCDD on palatal development in the embryos at 15.5 d of gestation [26]. Immunoreactivity of Runx2 and osteopontin in the palatine bone and MyoD and desmin in the palatine muscle was lower in the embryos exposed to TCDD than in the control embryos [26]. Immunoreactivity of the AhR and ER-α was localized in the normal palatine bone, but ER-α activity in the palatine bone was decreased in the exposed embryos [26]. Western blot analysis indicated that Runx2, MyoD, and desmin in the palate were downregulated in the embryos exposed to TCDD [26]. These results suggested that TCDD may impair palatal osteogenesis and myogenesis via the AhR.

These results suggested that lactational exposure to TCDD affects bone growth primarily by affecting vitamin D and osteoblast activation, resulting in impaired bone mineralization and shape. Lactational exposure to TCDD may also negatively affect bone matrix maturation, resulting in altered bone material properties. Fetal exposure to TCDD impairs osteogenesis of the palatine bone, causing a cleft palate.

Effects of Exposure to TCDD in Adulthood on Bones

Alaluusua et al. (1993) administered a single dose of TCDD to young adult TCDD-resistant male H/W rats and found smaller skulls and thinner incisors than in the control rats 16 weeks after exposure [12]. They suggested that the impairment of skull and teeth formation may have been associated with exposure to TCDD affecting vitamin A metabolism [12].

Focusing on the sensitivity to the effects of TCDD being related to the AhR, Jamsa et al. (2001) investigated the effects of exposure to TCDD on bones in L/E rats that were very sensitive to exposure to TCDD and in H/W rats that were resistant to the effects of TCDD [27]. TCDD was administered to each rat once a week for 20 weeks starting at 10 weeks old. TCDD only affected the bones in H/W rats when the total TCDD dose was 170 μg/kg, but a marked decrease in bone growth (e.g., decreased tibia size) occurred in the L/E rats at a TCDD dose of 1.7 μg/kg [27]. However, bone density in the epiphyseal region remained unchanged in the L/E rats [27]. The breaking force of the tibia from the H/W rats was decreased at a TCDD dose of 170 μg/kg and stiffness of the tibia from the H/W rats was decreased at a TCDD dose of 17 μg/kg [27]. The breaking force and stiffness of the tibia from the L/E rats were both decreased at a TCDD dose of 17 μg/kg [27]. These results suggested that the AhR may be involved in the effects of TCDD on bone growth, modeling, and mechanical strength.

Herlin et al. (2010) percutaneously administered TCDD at doses of 1–1000 ng/kg/day to 10-week-old female L/E and H/W rats once each week for 20 weeks (0.14–140 μg/kg of total doses) and quantified changes in bone shape, calcification, and biomechanical properties caused by long-term exposure to TCDD [28]. The results indicated that TCDD affected the geometrical and biomechanical parameters of the bone. The strongest responses were found in the trabecular bone area of the proximal tibia and the intracortical circumference of the tibial epiphysis [28]. The most marked difference (by a factor of ~49) between the two rat strains was for the cross-sectional area of the proximal tibia, indicating that the AhR plays a role in bone toxicity caused by TCDD [28].

Herlin et al. (2013) administered TCDD weekly for 10 weeks at a total dose of 200 μg/kg to AhR knockout and wild-type mice and used molecular biological methods to confirm that the AhR was involved in bone toxicity caused by TCDD [29]. The bone matrices were stiffer, the cortical bones were thinner and more porous, and the trabecular bone compartments were more compact in the wild-type mice exposed to TCDD than in the AhR knockout mice exposed to TCDD [29]. Exposure to TCDD also affected the expression of bone metabolism markers and osteogenesis-related genes, suggesting that an imbalance in bone remodeling occurred. In contrast, exposure to TCDD caused minimal bone changes in the knockout mice [29]. These results suggested that the AhR is involved in normal bone development and plays an important role in the osteo-toxic mechanism of TCDD.

Fader et al. (2018) investigated the effects of TCDD (at doses of 0.01–30 μg/kg) on the femurs of juvenile mice and found dose-dependent increases in the trabecular bone volume in both sexes [30]. Exposure to TCDD decreased serum tartrate-resistant acid phosphatase levels and osteoclast surface-to-bone surface ratios and inhibited femoral bone proteases, such as cathepsin K and matrix metallopeptidase 13, indicating that TCDD inhibited bone resorption [30]. Exposure to TCDD also increased the number of osteoblasts and decreased the number of bone marrow adipocytes on the trabecular bone surface, suggesting that the AhR may activate the differentiation of mesenchymal stem cells into osteoblasts. Analysis using an RNA sequencer indicated that transmembrane glycoprotein NMB expression and expression of a positive regulator of osteoblast differentiation and mineralization were dose-dependently induced by exposure to TCDD [30]. Exposure to TCDD also increased the 1,25-dihydroxy vitamin D3 concentration in serum, consistent with induction of 1α-hydroxylase Cyp27b1 in the kidneys, indicating that TCDD may have contributed to impaired bone resorption [30]. These results suggested that AhR activation by TCDD may shift bone remodeling toward bone formation and decrease the production of bone marrow fat.

Thus, the exposure of juveniles to TCDD altered bone shape, decreased bone size, and decreased the bone breaking force and stiffness, particularly of the tibia, by affecting bone remodeling via the AhR. However, TCDD was not found to affect bone density. Inconsistent positive effects of TCDD on bone remodeling have also been reported. TCDD exposure has been found to increase bone formation in the femur by inhibiting bone resorption via vitamin D activation and increased activation of osteoblast differentiation from mesenchymal stem cells. Further studies will be required to clarify the effects of TCDD on various bones and at lower TCDD doses.

3.2.2. Studies of Humans: Epidemiological and Clinical Studies

No epidemiological studies aimed at investigating associations between bone parameters and perinatal exposure to PCDD/Fs have been published. The human studies described here used populations mainly exposed to PCDD/Fs during childhood and adulthood (Table 4).

Akamine et al. (1985) found that alveolar bone resorption occurred in at least one tooth in 56.4% of Yusho patients and more commonly in men than women [31]. Severe alveolar bone resorption was found in patients in their 20 s and 30 s. Shimizu et al. (1992) monitored Yusho patients for 12 years and found that many patients had periodontal disease with horizontal alveolar bone resorption despite good oral care [32]. These clinical results suggested that mineralization of the alveolar bone may be affected by exposure to PCDD/Fs.

Hodgson et al. (2008) used dual-energy X-ray absorptiometry (DEXA) to measure the bone mineral densities (BMDs) of the forearm bones of 154 men and 167 women 60–81 years old living near a PCB-contaminated river on the Baltic Sea coast and also assessed the relationships between the BMDs and the concentrations of five dl-PCBs, three non-dl-PCBs, and p,p′-dichlorodiphenyldichloroethylene [33]. In men, the PCB-118 (a dl-PCB) concentration negatively correlated with the BMD, and the odds ratio was 1.06 (95% confidence intervals 1.01 and 1.12) for each increase of 10 pg/mL in the PCB-118 concentration for a low BMD (Z-score < -1) [33]. However, the sum of the concentrations of the three most abundant non-dl-PCB positively correlated with the BMD [33]. In women, the PCB-118 concentration positively correlated with the BMD and exposure to PCBs was not found to be a risk factor for decreased BMD [33].

Table 4. Effects of dioxins on bones in humans.

No.	Author	Publication Year	Country/Region	Target Population (Gender, Age, Residents/Patients)	Exposure Indicators	Bone Effect Markers
1	Akamine et al. [31]	1985	Japan	Yusho patients aged at 20 s to 60 s	dl-PCB and PCDF congeners in sera	Higher prevalence of sever alveolar bone resorption in patients compared with healthy subjects in the same age groups.
2	Shimizu et al. [32]	1992	Japan	Yusho patients	dl-PCB and PCDF congeners in sera	Increased morbidity of periodontal diseases with horizontal alveolar bone resorption during 12-year observation
3	Hodgson et al. [33]	2008	Sweden	Residents aged at 60–81 year (154 men and 167 women) living in Baltic Sea coast areas	Five dl-PCBs PCB118, three non-dl PCBs, and p,p′-DDE in blood	Low bone density correlated with PCB118 in men, but positive correlation between bone density and PCB118 in women.
4	Eskenazi et al. [7]	2014	Seveso, Italy	350 women who were under 20 years old in 1976	TCDD in sera taken immediately after the explosion	Bone density and size and strength indices of the three hip-forming bone regions. Better bone structure in highly exposed women, but no association between bone density and TCDD.
5	Fukushi et al. [34]	2016	Japan	Residents (262 women and 227 men) including Yusho patients (61.5% in women and 69.6% in men)	PCDDs, PCDFs, and non-ortho PCBs in blood	1,2,3,4,6,7,8-HpCDD may have a negative effect on bone mineral density in women, but serum levels of this conjugate were not increased in patients with Yusho.

Eskenazi et al. (2014) investigated the association between exposure to TCDD in youth and the bone structure and skeleton in adulthood [7]. They also investigated whether the timing of exposure to TCDD affected the association [7]. DEXA bone density measurements of the lumbar spine and hip and hip structural analysis measurements to determine size and strength indices for the three hip-forming bone regions were performed on 350 women in Seveso, Italy, who were <20 years old at the time of the Seveso accident [7]. The relationships between these bone indices and the TCDD concentrations in serum samples taken from the women immediately after the pesticide plant explosion were assessed [7]. The results indicated that the TCDD concentration was associated with several indicators of good bone structure in premenopausal women exposed to TCDD before peak bone mass had been reached [7]. The association was stronger for women exposed to TCDD before the age of 5 years old. Better bone structure was found for postmenopausal women exposed to TCDD after peak bone mass had been achieved than was found for postmenopausal women exposed to TCDD before peak bone mass was achieved [7]. These results suggested that exposure to TCDD may strengthen bones and not negatively affect bones in women before or after menopause.

Fukushi et al. (2016) assessed associations between PCDD, PCDF, and non-ortho-chlorinated biphenyl concentrations in serum and BMD determined by dual-energy X-ray absorptiometry in 262 female and 227 male Yusho patients [34]. After adjusting for area differences, the 1,2,3,4,7,8-hexachlorodibenzo-p-dioxin and 2,3,7,8-tetradibenzofuran concentrations were found to significantly positively correlate with the BMD Z-score for males [34]. No significant correlations were found between the PCDD, PCDF, and non-ortho-chlorinated biphenyl concentrations in serum and the BMDs for females. However, after adjusting for area and body mass index, the 1,2,3,4,6,7,8-heptachlorodibenzo-p-dioxin concentration was found to negatively correlate with the BMD Z-score for females [34]. These results indicated that exposure to some PCDD/F congeners may decrease bone density in women. Studies with more confounding factors controlled will be required to determine more clearly the effects of PCDD/Fs on bone density in Yusho patients.

No clear adverse effects of PCDD/Fs on bone density leading to osteoporosis were found in the epidemiological studies mentioned above, although altered mineralization of alveolar bone may have occurred in people exposed to high PCDD/F concentrations (e.g., Yusho patients). Better bone structure was associated with exposure to TCDD in women affected by the Seveso accident, but this may have been caused by TCDD causing bone remodeling to occur as bone formation, as was found in an animal study performed by Fader et al. (2018) [30].

4. Discussion

4.1. Effects of PCDD/Fs on Teeth and Their Potential Mechanisms

In animals, fetal and lactational exposure to TCDD affect the development of molars, resulting in defects, delayed eruption, and impaired calcification. The early stages of tooth morphogenesis in the fetal period are most sensitive to TCDD. In vitro studies of TCDD effects on the process of tooth formation using mouse molar embryos, it was reported that the impairment of epidermal growth factor receptor (EGFR) [35] and promoting epithelial apoptosis [36] may play important roles in the toxic effects of TCDD in fetuses. Moreover, Kiukkonen et al. (2006) investigated the associations between impaired mineralization of mandibular molars in mouse embryo budding stage of E18 by TCDD and expression of dentin sialophosphoprotein (Dspp), Bono1, and matrix metalloproteinase-20 (MMP-20), which are involved in hard tissue calcification [37]. Although no clear differences in the localization and intensity of Bono1 and MMP-20 expression were observed, Dspp expression in secretory odontoblasts and in presecretory ameloblasts was reduced in embryos incubated with TCDD. These findings suggest that TCDD may cause impaired calcification of dentin due to toxicity specific to Dspp expression.

Exposure to relatively high TCDD doses in adulthood can affect both mesenchymal and epithelial elements in the incisors, resulting in cessation of dentin formation and qualitative changes in the enamel [13]. However, no difference of impaired formation of the incisor tooth was found between rat strains with different AhR sensitivities [13], suggesting a pathway other than AhR signaling might mediate the effects of high TCDD exposure on teeth in adult rats.

In human studies, delayed tooth eruption and broken teeth occurred in Yusho patients exposed to high PCB, PCQ, and PCDF concentrations. Enamel dysplasia has been suggested to be a characteristic symptom in Finnish children exposed to high PCDD/F concentrations, and residents of Seveso, Italy, exposed to extremely high TCDD concentrations. However, associations between PCDD/F concentrations and tooth growth, including enamel dysplasia, have been investigated only in three studies [4–6] with populations whose perinatal exposure to PCDD/Fs was quantified, one of which showed no association between PCDD/Fs and growth of molars. More studies of perinatally exposed populations will be required to clarify the effects of exposure to PCDD/Fs on teeth.

4.2. Effects of PCDD/Fs on Bone and Their Potential Mechanisms

In animal studies, lactational exposure to TCDD was found to affect vitamin D and osteoblast activation in the tibia [24], resulting in impaired bone mineralization and shape and altered bone material properties. Fetal exposure to TCDD impaired osteogenesis and myogenesis of the palatine by decreasing Runx2 and osteopontin in the palatine bone and MyoD and desmin in the palatine muscle, leading to cleft palates [26].

In juvenile rats exposed to TCDD, the tibia shape was altered, and the size, breaking force, and stiffness were decreased but no effect of TCDD on bone density was found. However, exposure to TCDD at low doses increased bone formation in femurs by inhibiting bone resorption via vitamin D activation and increased activation of osteoblast differentiation from mesenchymal stem cells [30].

In an in vitro study using a system in which stem cells were isolated from the bone marrow of rat and mouse femurs and tibias, Korkalainen et al. (2009) investigated the effects of TCDD exposure including low dose of TCDD, e.g., 100 fM, on differentiation of osteoblasts and osteoclasts [38]. During osteoblast differentiation, TCDD significantly and dose-dependently decreased mRNA levels of RUNX2, alkaline phosphatase, and osteocalcin. In the case of osteoclasts, TCDD also reduced the number of TRACP+ multinucleated cells, with a concomitant decrease in the number and resorption area of F-actin rings. Effects on osteoblasts and osteoclasts occurred at very low doses of TCDD exposure. Taken together, the disturbance of osteoblast differentiation may also affect osteoclast formation resulting in the qualitative changes in bone caused by TCDD observed in in vivo studies.

Liu X et al. (2020) firstly confirmed that TCDD decreased cell proliferation and calcium deposition of the osteoblasts from human fetal palatal mesenchymal cells [39]. Then, they investigated molecular mechanisms of TCDD-induced inhibition of these osteogenic cells and reported that osteogenic cell differentiation was inhibited by downregulation of BMP-2/TGF-β1/Smad pathway via AhR signaling, suggesting that crosstalk between AhR and BMP-2/TGF-β1/Smad signaling may have an important role in osteogenic impairment of fetal palatal bone by TCDD [39].

In humans, no clear adverse effects of PCDD/Fs on bone were found in epidemiological studies using bone density as an indicator of osteoporosis. Better bone structures were associated with exposure to TCDD in women exposed to high TCDD concentrations in Seveso, Italy. However, this may have been caused by TCDD causing bone remodeling to occur in the direction of bone formation, as was found in an animal study.

5. Conclusions

Fetal exposure to PCDD/Fs appears to alter odontogenesis, particularly enamel formation, but has a limited effect on only palatine bone genesis. Lactational exposure to PCDD/Fs may affect both teeth and bones, but the effects on bones may be reversible. Exposure to high PCDD/F concentrations may affect teeth even during adulthood, but exposure to PCDD/Fs may induce osteogenesis and improve bone properties because PCDD/Fs may disrupt bone remodeling. Contrary to bones, teeth, and particularly the enamel of teeth, are not remodeled after birth, suggesting that developmental enamel defects can be used as biomarkers of PCDD/Fs exposure.

More epidemiological studies of humans are required to clarify associations between exposure to PCDD/Fs and biological markers of teeth and bones (including vitamin D metabolites) in the future.

Author Contributions: Conceptualization, T.T. and Y.N.; writing—original draft preparation, T.T. and H.T.V.; writing—review and editing, T.T. and H.T.V. All authors have read and agreed to the published version of the manuscript.

Funding: This research received no external funding.

Institutional Review Board Statement: Not applicable.

Informed Consent Statement: Not applicable.

Data Availability Statement: Not applicable.

Acknowledgments: We thank Gareth Thomas for editing a draft of this manuscript.

Conflicts of Interest: The authors declare no conflict of interest.

References

1. Suskind, R.R.; Hertzberg, V.S. Human health effects of 2,4,5-T and its toxic contaminants. *JAMA* **1984**, *251*, 2372–2380. [CrossRef] [PubMed]
2. Yamashita, F.; Hayashi, M. Fetal PCB syndrome: Clinical features, intrauterine growth retardation and possible alteration in calcium metabolism. *Environ. Health Perspect.* **1985**, *59*, 41–45. [PubMed]
3. Alaluusua, S.; Lukinmaa, P.L.; Torppa, J.; Tuomisto, J.; Vartiainen, T. Developing teeth as biomarker of dioxin exposure. *Lancet* **1999**, *353*, 206. [CrossRef] [PubMed]

4. Alaluusua, S.; Lukinmaa, P.L.; Vartiainen, T.; Partanen, M.; Torppa, J.; Tuomisto, J. Polychlorinated dibenzo-p-dioxins and dibenzofurans via mother's milk may cause developmental defects in the child's teeth. *Environ. Toxicol. Pharmacol.* **1996**, *1*, 193–197. [CrossRef]
5. Laisi, S.; Kiviranta, H.; Lukinmaa, P.L.; Vartiainen, T.; Alaluusua, S. Molar-incisor-hypomineralisation and dioxins: New findings. *Eur. Arch. Paediatr. Dent.* **2008**, *9*, 224–227. [CrossRef]
6. Alaluusua, S.; Calderara, P.; Gerthoux, P.M.; Lukinmaa, P.L.; Kovero, O.; Needham, L.; Patterson, D.G., Jr.; Tuomisto, J.; Mocarelli, P. Developmental dental aberrations after the dioxin accident in Seveso. *Environ. Health Perspect.* **2004**, *112*, 1313–1318. [CrossRef]
7. Eskenazi, B.; Warner, M.; Sirtori, M.; Fuerst, T.; Rauch, S.A.; Brambilla, P.; Mocarelli, P.; Rubinacci, A. Serum dioxin concentrations and bone density and structure in the Seveso Women's Health Study. *Environ. Health Perspect.* **2014**, *122*, 51–57. [CrossRef]
8. Kattainen, H.; Tuukkanen, J.; Simanainen, U.; Tuomisto, J.T.; Kovero, O.; Lukinmaa, P.L.; Alaluusua, S.; Tuomisto, J.; Viluksela, M. In utero/lactational 2,3,7,8-tetrachlorodibenzo-p-dioxin exposure impairs molar tooth development in rats. *Toxicol. Appl. Pharmacol.* **2001**, *174*, 216–224. [CrossRef]
9. Lukinmaa, P.L.; Sahlberg, C.; Leppaniemi, A.; Partanen, A.M.; Kovero, O.; Pohjanvirta, R.; Tuomisto, J.; Alaluusua, S. Arrest of rat molar tooth development by lactational exposure to 2,3,7,8-tetrachlorodibenzo-p-dioxin. *Toxicol. Appl. Pharmacol.* **2001**, *173*, 38–47. [CrossRef]
10. Miettinen, H.M.; Alaluusua, S.; Tuomisto, J.; Viluksela, M. Effect of in utero and lactational 2,3,7,8-tetrachlorodibenzo-p-dioxin exposure on rat molar development: The role of exposure time. *Toxicol. Appl. Pharmacol.* **2002**, *184*, 57–66. [CrossRef]
11. Gao, Y.; Sahlberg, C.; Kiukkonen, A.; Alaluusua, S.; Pohjanvirta, R.; Tuomisto, J.; Lukinmaa, P.L. Lactational exposure of Han/Wistar rats to 2,3,7,8-tetrachlorodibenzo-p-dioxin interferes with enamel maturation and retards dentin mineralization. *J. Dent. Res.* **2004**, *83*, 139–144. [CrossRef] [PubMed]
12. Alaluusua, S.; Lukinmaa, P.L.; Pohjanvirta, R.; Unkila, M.; Tuomisto, J. Exposure to 2,3,7,8-tetrachlorodibenzo-para-dioxin leads to defective dentin formation and pulpal perforation in rat incisor tooth. *Toxicology* **1993**, *81*, 1–13. [CrossRef] [PubMed]
13. Kiukkonen, A.; Viluksela, M.; Sahlberg, C.; Alaluusua, S.; Tuomisto, J.T.; Tuomisto, J.; Lukinmaa, P.L. Response of the incisor tooth to 2,3,7,8-tetrachlorodibenzo-p-dioxin in a dioxin-resistant and a dioxin-sensitive rat strain. *Toxicol. Sci.* **2002**, *69*, 482–489. [CrossRef]
14. Miller, R.W. Congenital PCB poisoning: A reevaluation. *Environ. Health Perspect.* **1985**, *60*, 211–214. [CrossRef] [PubMed]
15. Rogan, W.J.; Gladen, B.C.; Hung, K.L.; Koong, S.L.; Shih, L.Y.; Taylor, J.S.; Wu, Y.C.; Yang, D.; Ragan, N.B.; Hsu, C.C. Congenital poisoning by polychlorinated biphenyls and their contaminants in Taiwan. *Science* **1988**, *241*, 334–336. [CrossRef]
16. Gladen, B.C.; Taylor, J.S.; Wu, Y.C.; Ragan, N.B.; Rogan, W.J.; Hsu, C.C. Dermatological findings in children exposed transplacentally to heat-degraded polychlorinated biphenyls in Taiwan. *Br. J. Dermatol.* **1990**, *122*, 799–808. [CrossRef]
17. Guo, Y.L.; Lambert, G.H.; Hsu, C.C. Growth abnormalities in the population exposed in utero and early postnatally to polychlorinated biphenyls and dibenzofurans. *Environ. Health Perspect.* **1995**, *103* (Suppl. 6), 117–122.
18. Alaluusua, S.; Kiviranta, H.; Leppaniemi, A.; Holtta, P.; Lukinmaa, P.L.; Lope, L.; Jarvenpaa, A.L.; Renlund, M.; Toppari, J.; Virtanen, H.; et al. Natal and neonatal teeth in relation to environmental toxicants. *Pediatr. Res.* **2002**, *52*, 652–655. [CrossRef]
19. Fukuyama, H.; Anan, Y.; Akamine, A.; Aono, M. [Alteration in stomatological findings of patients with yusho (PCB poisoning) in the general examination (author's transl)]. *Fukuoka Igaku Zasshi* **1979**, *70*, 187–198.
20. Guo, Y.L.; Yu, M.L.; Hsu, C.C.; Rogan, W.J. Chloracne, goiter, arthritis, and anemia after polychlorinated biphenyl poisoning: 14-year follow-Up of the Taiwan Yucheng cohort. *Environ. Health Perspect.* **1999**, *107*, 715–719. [CrossRef]
21. Kanagawa, Y.; Matsumoto, S.; Koike, S.; Tajima, B.; Fukiwake, N.; Shibata, S.; Uchi, H.; Furue, M.; Imamura, T. Association of clinical findings in Yusho patients with serum concentrations of polychlorinated biphenyls, polychlorinated quarterphenyls and 2,3,4,7,8-pentachlorodibenzofuran more than 30 years after the poisoning event. *Environ. Health* **2008**, *7*, 47. [CrossRef] [PubMed]
22. Ngoc, V.T.N.; Huong, L.T.; Van Nhon, B.; Tan, N.T.M.; Van Thuc, P.; Hien, V.T.T.; Dung, T.M.; Van Toan, N.; Anh, L.Q.; Son, L.H.; et al. The higher prevalence of developmental defects of enamel in the dioxin-affected region than non-dioxin-affected region: Result from a cross-sectional study in Vietnam. *Odontology* **2019**, *107*, 17–22. [CrossRef] [PubMed]
23. Miettinen, H.M.; Pulkkinen, P.; Jamsa, T.; Koistinen, J.; Simanainen, U.; Tuomisto, J.; Tuukkanen, J.; Viluksela, M. Effects of in utero and lactational TCDD exposure on bone development in differentially sensitive rat lines. *Toxicol. Sci.* **2005**, *85*, 1003–1012. [CrossRef] [PubMed]
24. Nishimura, N.; Nishimura, H.; Ito, T.; Miyata, C.; Izumi, K.; Fujimaki, H.; Matsumura, F. Dioxin-induced up-regulation of the active form of vitamin D is the main cause for its inhibitory action on osteoblast activities, leading to developmental bone toxicity. *Toxicol. Appl. Pharmacol.* **2009**, *236*, 301–309. [CrossRef] [PubMed]
25. Finnila, M.A.; Zioupos, P.; Herlin, M.; Miettinen, H.M.; Simanainen, U.; Hakansson, H.; Tuukkanen, J.; Viluksela, M.; Jamsa, T. Effects of 2,3,7,8-tetrachlorodibenzo-p-dioxin exposure on bone material properties. *J. Biomech.* **2010**, *43*, 1097–1103. [CrossRef] [PubMed]
26. Yamada, T.; Hirata, A.; Sasabe, E.; Yoshimura, T.; Ohno, S.; Kitamura, N.; Yamamoto, T. TCDD disrupts posterior palatogenesis and causes cleft palate. *J. Craniomaxillofac. Surg.* **2014**, *42*, 1–6. [CrossRef] [PubMed]
27. Jamsa, T.; Viluksela, M.; Tuomisto, J.T.; Tuomisto, J.; Tuukkanen, J. Effects of 2,3,7,8-tetrachlorodibenzo-p-dioxin on bone in two rat strains with different aryl hydrocarbon receptor structures. *J. Bone Miner. Res.* **2001**, *16*, 1812–1820. [CrossRef]
28. Herlin, M.; Kalantari, F.; Stern, N.; Sand, S.; Larsson, S.; Viluksela, M.; Tuomisto, J.T.; Tuomisto, J.; Tuukkanen, J.; Jamsa, T.; et al. Quantitative characterization of changes in bone geometry, mineral density and biomechanical properties in two rat strains

with different Ah-receptor structures after long-term exposure to 2,3,7,8-tetrachlorodibenzo-p-dioxin. *Toxicology* **2010**, *273*, 1–11. [CrossRef]

29. Herlin, M.; Finnila, M.A.; Zioupos, P.; Aula, A.; Risteli, J.; Miettinen, H.M.; Jamsa, T.; Tuukkanen, J.; Korkalainen, M.; Hakansson, H.; et al. New insights to the role of aryl hydrocarbon receptor in bone phenotype and in dioxin-induced modulation of bone microarchitecture and material properties. *Toxicol. Appl. Pharmacol.* **2013**, *273*, 219–226. [CrossRef]
30. Fader, K.A.; Nault, R.; Raehtz, S.; McCabe, L.R.; Zacharewski, T.R. 2,3,7,8-Tetrachlorodibenzo-p-dioxin dose-dependently increases bone mass and decreases marrow adiposity in juvenile mice. *Toxicol. Appl. Pharmacol.* **2018**, *348*, 85–98. [CrossRef]
31. Akamine, A.; Hashiguchi, I.; Maeda, K.; Hara, Y.; Chinjiu, N.; Iwamoto, Y.; Aono, M. [Prevalence of periodontal disease in patients with yusho]. *Fukuoka Igaku Zasshi* **1985**, *76*, 248–252. [PubMed]
32. Shimizu, K.; Nakata, S.; Murakami, T.; Tamari, K.; Takahama, Y.; Akamine, A.; Aono, M. Long-term occlusal guidance of a severely intoxicated patient with yusho (PCB poisoning): A case report. *Am. J. Orthod. Dentofacial. Orthop.* **1992**, *101*, 393–402. [CrossRef] [PubMed]
33. Hodgson, S.; Thomas, L.; Fattore, E.; Lind, P.M.; Alfven, T.; Hellstrom, L.; Hakansson, H.; Carubelli, G.; Fanelli, R.; Jarup, L. Bone mineral density changes in relation to environmental PCB exposure. *Environ. Health Perspect.* **2008**, *116*, 1162–1166. [CrossRef] [PubMed]
34. Fukushi, J.; Tokunaga, S.; Nakashima, Y.; Motomura, G.; Mitoma, C.; Uchi, H.; Furue, M.; Iwamoto, Y. Effects of dioxin-related compounds on bone mineral density in patients affected by the Yusho incident. *Chemosphere* **2016**, *145*, 25–33. [CrossRef] [PubMed]
35. Partanen, A.M.; Alaluusua, S.; Miettinen, P.J.; Thesleff, I.; Tuomisto, J.; Pohjanvirta, R.; Lukinmaa, P.L. Epidermal growth factor receptor as a mediator of developmental toxicity of dioxin in mouse embryonic teeth. *Lab. Invest.* **1998**, *78*, 1473–1481.
36. Partanen, A.M.; Kiukkonen, A.; Sahlberg, C.; Alaluusua, S.; Thesleff, I.; Pohjanvirta, R.; Lukinmaa, P.L. Developmental toxicity of dioxin to mouse embryonic teeth in vitro: Arrest of tooth morphogenesis involves stimulation of apoptotic program in the dental epithelium. *Toxicol. Appl. Pharmacol.* **2004**, *194*, 24–33. [CrossRef]
37. Kiukkonen, A.; Sahlberg, C.; Lukinmaa, P.L.; Alaluusua, S.; Peltonen, E.; Partanen, A.M. 2,3,7,8-tetrachlorodibenzo-p-dioxin specifically reduces mRNA for the mineralization-related dentin sialophosphoprotein in cultured mouse embryonic molar teeth. *Toxicol. Appl. Pharmacol.* **2006**, *216*, 399–406. [CrossRef]
38. Korkalainen, M.; Kallio, E.; Olkku, A.; Nelo, K.; Ilvesaro, J.; Tuukkanen, J.; Mahonen, A.; Viluksela, M. Dioxins interfere with differentiation of osteoblasts and osteoclasts. *Bone* **2009**, *44*, 1134–1142. [CrossRef]
39. Liu, X.; Li, X.; Tao, Y.; Li, N.; Ji, M.; Zhang, X.; Chen, Y.; He, Z.; Yu, K.; Yu, Z. TCDD inhibited the osteogenic differentiation of human fetal palatal mesenchymal cells through AhR and BMP-2/TGF-beta/Smad signaling. *Toxicology* **2020**, *431*, 152353. [CrossRef]

toxics MDPI

Systematic Review

Research on the Relationship between Exposure to Dioxins and Cancer Incidence in Vietnam

Tuong Phi Vuong

Investigation Center No 69, Hanoi 118700, Vietnam; tphivuong@yahoo.com

Abstract: The aim of this literature review is to discover whether there is a relationship between exposure to dioxins and cancer incidence in the hotspot regions of Vietnam by estimating the risk ratio index. The results of the study show that the incidence of cancer (soft tissue sarcoma; Hodgkin's and non-Hodgkin's lymphoma; lung, prostate, and liver cancer) in the dioxin-exposed Vietnamese population is much higher than the results of studies published in other countries because of the high levels of dioxins in South Vietnam, where Agent Orange was sprayed during the war. Further studies on the health effects of dioxins in the Vietnamese population, including cancer incidence, should be conducted with improved research methods.

Keywords: dioxins; cancer; incidence; risk ratio; Vietnam

Citation: Vuong, T.P. Research on the Relationship between Exposure to Dioxins and Cancer Incidence in Vietnam. *Toxics* **2022**, *10*, 384. https://doi.org/10.3390/toxics10070384

Academic Editor: Muneko Nishijo

Received: 24 March 2022
Accepted: 6 July 2022
Published: 11 July 2022

1. Introduction

Dioxin is a general term that describes a group of hundreds of toxic organochlorine compounds that are highly persistent in the environment. The most toxic compound is 2,3,7,8-tetrachlorodibenzo-p-dioxin or TCDD. This dioxin compound was classified by the International Organization for Research on Cancer (IARC) as a human carcinogen (Group 1) based on animal studies, novel knowledge about its cytotoxicity mechanisms [1], and human studies on occupational exposure [2].

Dioxins have caused ecological disasters in Seveso (Italy) [3], Times Beach (Missouri), and the Love Canal (New York), but the most severe and long-lasting contamination occurred in South Vietnam, where Agent Orange was sprayed by the US for 10 years [4].

There are numerous epidemiological studies concerning the relationship between dioxins and cancer development in humans using different statistical indexes such as the odds ratio (OR), risk ratio (RR), standard incidence ratio (SIR), and standardized mortality ratio (SMR) for the cancer incidence or mortality of populations exposed to dioxins. However, epidemiological data on the carcinogenic potential of dioxins in Vietnamese people still have certain limitations because they have only been reported in observational epidemiological studies. Thus, the aim of this literature review is to provide a quantitative assessment of the association from an epidemiological point of view by estimating the risk ratio (RR) of cancer incidence in Vietnamese people exposed to dioxins.

2. Materials and Methods

2.1. Data Sources, Search Strategy and Selection Criteria

Literature searches were conducted in Google Scholar in both English and Vietnamese to identify eligible studies. The following terms were used in the search procedure: ("dioxin" or "TCDD" or "Agent Orange) AND ("cancer incidence") AND ("Vietnamese" or "Vietnam").

Studies were eligible for inclusion if all of the following criteria were fulfilled: (1) the studies evaluated the association between dioxin/TCDD and cancer incidence; (2) the risk ratio (RR) or odds ratio (OR) estimates were available for evaluation; (3) the articles as full papers were written in English or Vietnamese. (4) Reviews, meeting abstracts, notes, comments, editorials, and case reports were excluded because of their limited data.

2.2. Data Extraction and Synthesis

For the English literature review, we concentrated on published studies conducting a meta-analysis and official literature concerned with dioxins and cancer. The following information was extracted from each study: author(s), year of publication, country of study, study period, population characteristics (sample size, gender and age), and cancer subtypes risk ratios.

The Vietnamese literature review was followed by a data analysis to assess the association between dioxin exposure and all cancer incidences. For studies covering the relationship between exposure to dioxins and human cancer development, the risk ratio was estimated from the cumulative incidence:

$$\text{Risk Ratio} = CI_e / CI_u$$

where CI_e is the cumulative incidence in the 'exposed' group, and CI_u is the cumulative incidence in the 'unexposed' group. A 95% confidence interval (95% CI) was calculated.

3. Results

3.1. Dioxins and Cancer around the World

The latest publication from the IARC was published in 2012 [2] and comprehensively reviews epidemiological studies for any relationships between dioxins and human cancer development. Many studies have been carried out in the US [5–9], Germany [10,11], and Italy [12], and there have also been some multinational studies [13], including many cohort and case–control studies, on very large numbers of participants exposed to dioxins, mainly due to occupational exposure or environmental pollution incidents.

These studies were then analyzed by Xu et al. in 2016 in a systemic review [14], which involved 18,969 cancer cases and 3,155,159 participants included in 5 cohort studies and 5 case–control studies to assess the association between external exposure to dioxins and cancer incidence. Six of the studies with a cancer incidence RR are described in more detail in Table 1.

Table 1. Characteristics and results (RR) of some of the epidemiological studies on relationships between dioxins and human cancer development. Reproduced from Xu et al. [14], published by *Sci. Rep.*, 2016.

No.	Study	Country/ Cohort	Time Period	Exposure Way	Exposure Assessment	Cancer Types	Gender	No. of Cancer Cases/Cohort or Controls	Age (Years)	RR
External exposure to TCDD and cancer incidence (exposure incidence)										
1	Kogevinas [15]	Part of IARC	1955–1988	Occupational	Job and company records, questionnaires	All cancers, breast cancer	F	29/701	N/A	2.22
2	Read [16]	New Zealand	1970–2001	Nonoccupational	Individual records	All cancers	F/M	8013/375,583	N/A	0.91–1.11
3	Pesatori [3]	Italy, Seveso	1977–1996	Industrial accident	Measurements of TCDD soil levels	All cancers	F/M	2122/218,761	0–74	Zone A: 1.03 Zone B: 1.00 Zone R: 0.96
Blood level of TCDD and cancer incidence (blood incidence)										
4	Pavuk [8]	USA, Vietnam Veterans	1982–2003	Vietnam war	Physical examinations and blood samples	All cancers	M	402/1482	Mean 63.7	1.6
5	Warner [12]	Italy, Seveso, SWHS cohort	I:1976–1996, II:1997–2009	Industrial accidents	Interviews, physical and blood samples, and sample examination	All cancers, breast cancer	F	66/981	0–40	2.77
6	Ott [11]	Germany, Ludwigshafe	1959–1992	Occupational	Questionnaires and blood samples	All cancers	M	31/243	N/A	1.3

The pooled RR of the all-cancer incidence with external exposure to TCDD was 1.01 (95% CI: 0.97–1.06), indicating no significant association between external exposure to TCDD and cancer incidence. There was significant heterogeneity across the included studies (I2 = 73.5%, $p < 0.001$). A subgroup analysis was conducted according to cancer subtypes (Table 2).

Table 2. Analysis of the association between external exposure to TCDD and cancer incidence. Reproduced from Xu et al. [14], published by *Sci. Rep.*, 2016.

Type of Cancer	Pooled Risk Ratio (RR)	95% CI
All cancers	1.01	0.97–1.06
Soft tissue sarcoma	1.37	0.97–1.93
Hodgkin's Lymphoma	1.13	0.83–1.54
Non-Hodgkin's Lymphoma	1.09	0.92–1.30

All of the pooled RRs of exposure incidence cancer subtypes were not significant, including those for Hodgkin's lymphoma, non-Hodgkin's lymphoma, and soft tissue sarcoma (Table 2). Other cancer subtypes were also studied, but there was no information about their exposure incidence risk ratios.

One of the most important results of the study by Xu et al. [14] was the evidence of a significant positive association between cancer incidence, not only with external exposure, but also with blood TCDD level (the pooled RR for all cancer incidence was 1.57 (95% CI: 1.21–2.04)) (data not shown in Table 1). The subgroup analysis of blood incidence was not conducted due to the limited data.

Based on the literature review of prior studies such as Veterans and Agent Orange: Update 11 (2018) [17] and different annual Agent Orange and dioxin studies, committee updates from the Vietnam Veterans of America determined that no other changes in the association level between the relevant exposure levels and other cancer types are currently observed, as there are either no published studies, or new evidence has supported the findings of earlier updates. Thus, the current findings on cancer can be summarized as follows: there is sufficient evidence of an association between dioxins and soft tissue sarcomas, Hodgkin's lymphoma, and non-Hodgkin's lymphoma; there is limited or suggestive evidence of an association between dioxins and bladder cancer, laryngeal cancer, cancers of the lung, bronchus, and trachea, prostate cancer, multiple myeloma, and AL amyloidosis; there is inadequate or insufficient evidence to determine whether there is an association between dioxins and any other specific type of cancer.

In the 2020 issue of the annual Agent Orange Newsletter [18], the US Department of Veterans Affairs presumed that prostate cancer, respiratory cancers, chronic B-cell leukemias, and multiple myeloma are related to herbicide exposure, along with soft tissue sarcomas and Hodgkin's and non-Hodgkin's lymphomas. However, more clear evidence is needed to further epidemiological studies.

3.2. Dioxins and Cancer in Vietnam

Research by Kahn (1988) [19], Schecter (1990) [20], and Michalek (1990) [7] shows that the concentration of dioxins stored in the adipose tissue in the bodies of the US veterans who participated in the Vietnam War was 600 times higher than normal (600 ppt vs. 1–2 ppt) after many years. The TCDD geographic distribution study showed that TCDD rates were higher in the Vietnamese population in the southern and central provinces [21]. TCDD concentrations ranged from non-detectable to 1648 pg/g of lipids in human biological samples obtained from 1970 to 2017, while that of US adults was only 5.2 pg/g (measured by the US National Health and Nutrition Examination Survey in 2003–2004). The highest dioxin concentration recorded was in a fisherman at the lake in the Da Nang Air Base, which was over 1000 ppt [22]. In addition to an increase in the dioxin levels in the bodies of the Vietnamese people, dioxin levels also increased in the food and wildlife in these contaminated areas. Dioxin has been found in Vietnamese soil or sediments at concentrations of up to 1 million ppt 30–40 years after the areas had been sprayed with Agent Orange. The lipid-adjusted dioxin toxic equivalent (TEQ) in poultry in Vietnam was 25.8 pg/g compared to 0.018 pg/g for the US samples (Lorber et al., 2009) [23]. The mean TCDD concentration reported in eggs from Vietnam was approximately 4 pg/g, an order of magnitude higher than that from the US (in California) in the late 1990s (Goldman et al., 2000) [24].

Research on diseases related to toxic chemicals/dioxins in Vietnam veterans by Le et al. [25] was conducted in the period between 1994–2004, in which the patient files of and structured questionnaires administered to 47,893 veterans in eight provinces and cities across the country who had participated in the Vietnam War, which took place from 1962 to April 1975, were analyzed. The cancer incidence in the group with a history of exposure to toxic chemicals/dioxins was statistically significantly higher than in the unexposed group ($p < 0.01$). The incidence of lung cancer was 1.30%; Hodgkin's lymphoma—0.20%; non-Hodgkin's lymphoma—0.49%; sarcoma—0.19%; and prostate cancer—0.57%. All of these cancer subtypes have been recognized by the American Academy of Sciences as being related to toxic chemicals/dioxins. Based on the results that liver cancer incidence in exposed veterans was 0.93%, the authors also proposed adding liver cancer to the list of cancers related to toxic chemicals/dioxins. For a better understanding of the effects of dioxins on cancer incidence in exposed veterans, we calculated the risk ratio (RR) for each type of cancer in the study based on the study data. The results are presented in Table 3 below.

Table 3. Risk ratio (RR) of cancer among Vietnamese veterans exposed to dioxins in the years 1962–1975.

	Exposed Group *N* = 28,817		Unexposed Group *N* = 19,076		Risk Ratio	
Diseases	**Number**	**Incidence (%)**	**Number**	**Incidence (%)**	**RR**	**95% CI**
All cancers	1262	4.38	220	1.15	3.80	3.29–4.38
Soft tissue sarcoma	54	0.19	5	0.03	7.15	2.86–17.87
Hodgkin's Lymphoma	58	0.20	3	0.02	12.80	4.01–40.83
Non-Hodgkin's Lymphoma	142	0.49	29	0.15	3.24	2.17–4.83
Carcinomas of the lung, bronchus, and trachea.	577	2.00	111	0.58	3.44	2.81–4.21
Prostate cancer	164	0.57	37	0.19	2.93	2.05–4.19
Liver cancer	267	0.93	35	0.18	5.05	3.55–7.18

Note: Data from Le et al. (2007) [25].

The results of the risk ratio calculations showed that cancer incidence in exposed Vietnamese veterans is higher than that in unexposed groups, with a range that is 3.24 times (in respiratory cancer) to 12.8 times higher (in Hodgkin's lymphoma). The risk ratio for all cancer subtypes was 3.8 (95% CI: 3.29–4.38, $p < 0.0001$).

Do et al. in 2009 [26] also conducted research on the relationship between Agent Orange and cancer rates in Vietnamese people 30 years after the war. The study used data on cancer status that were either self-reported by the respondents or provided by others obtained from the Vietnam National Health Survey (N = 158,019) combined with data on military herbicide exposure obtained from information on the military activities of the US and allied troops during the war. Self-reported and proxy-reported cancer status were collected from the 2001–2002 Vietnam National Health Survey (VNHS). A stratified random sample of 1200 communes (out of a total of 8926 communes) and 36,000 households was interviewed from November 2001 to November 2002. The VNHS gathers individual data on self and proxy-reported morbidity. More precisely, the survey instrument asks the main respondent about all household members' health conditions. In addition to the VNHS, measurements of military herbicide exposure are constructed using the US Military Assistance Command Data Management Agency's Herbicide Report System (HERBS) file. A geographic information system (GIS) was further developed to facilitate the estimation of exposure from HERBS file data. Records combine the Combat Activities file, the South-East Asia Database, and the Combat Naval Gunfire File (CONGA) and cover the combat activities of allied forces over the period between October 1965 and June 1975 with the exception of a few missing months. Additional information on agricultural herbicide use, cigarette and alcohol consumption, and coal and firewood expenditures were obtained from the 2004 Vietnam Household Living Standards Survey to look at household-level confounding factors. The multivariate estimation method was used to analyze the data.

The study results did not show a significant difference in cancer rates between the communes infected with the military herbicide and the uninfected communes. Meanwhile, when only focusing on communes with toxic contamination, they found a positive relationship between exposure and cancer rates reported in 2001–2002. The results suggest that among communes that were previously exposed, an increase of 10 percent in herbicide exposure is associated with a 2 percent increase in the probability of reporting suffering from cancer.

In 2021, Nguyen et al. [27] conducted a study on the health status of Agent Orange victims in some southern provinces of Vietnam with 1675 participants, including 1107 men and 568 women with an average age of 61.9 years, by means of physical examination combined with structured questionnaire. The results are summarized in Table 4.

Table 4. Numbers and incidence of cancer among residents exposed to Agent Orange living in some of the provinces of South Vietnam.

Type of Cancer	Number	Incidence (%)
All cancers	163	9.73
Lung cancer	17	1.01
Liver cancer	23	1.37
Kidney cancer	34	2.03
Prostate cancer	69	6.23
Ovary cancer	7	1.23
Uterine cancer	13	2.29

Note: Data from Nguyen et al. (2021) [27] with 1675 participants including 1107 men and 568 women.

However, there was no control group that was unexposed to dioxins in this study. Meanwhile, we could not compare the obtained cancer incidence with the national cancer rate because the study was carried out on a particular population (mainly the elderly with an average age of 61.9 years). Evaluating the risk ratio of cancer subtypes by using data from the study by Le et al. [25] with the incidence of the unexposed group (19,076 participants) also could not be conducted because of a lack of information on gender rate in the unexposed group.

4. Discussion

Epidemiological studies on the relationship between dioxins and cancer incidence in Vietnamese people are relatively few. Using a search strategy and selection criteria, we found only three studies. Based on the results, we have calculated the risk ratios and compared them to those of other studies in other countries around the world.

The results show that the risk ratios of all cancers and different cancer subtypes in Vietnamese people exposed to dioxins (with high levels in the contaminated areas of Vietnam caused by the US army during the war) were higher (3.76 times for all cancers; 5.22 times for soft tissue sarcoma; 11.33 times for Hodgkin's; and 2.97 times for non-Hodgkin's lymphomas) than those in other countries where the examined people were exposed to lower amounts of dioxins (mainly caused by occupational exposure or environmental pollution incidents) (Table 5).

Table 5. Comparison of cancer incidence risk ratio for Vietnamese people with people from other countries exposed to dioxins.

Type of Cancer	Risk Ratio in Dioxin-Exposed Vietnamese in the Study of Le et al. *	Risk Ratio in Dioxin-Exposed Patient from other Countries **	$RR_{Vietnam}/RR_{Other\ Countries}$
All cancers	3.80	1.01	3.76
Soft tissue sarcoma	7.15	1.37	5.22
Hodgkin's Lymphoma	12.80	1.13	11.33
Non- Hodgkin's Lymphoma	3.24	1.09	2.97
Lung cancer	3.44 (4.40)	No data	-
Prostate cancer	2.93 (21.24)	No data	-
Liver cancer	5.05 (7.48)	No data	-

Note: Data (*) from study of Le et al. (2007) [25], data in parentheses from Nguyen et al. (2021) [27], and (**) from Xu et al. (2016) [14].

Despite an increased risk ratio for lung, prostate, and liver cancer in the dioxin-exposed Vietnamese people in our study, there is still only limited or suggestive evidence of an association between dioxins and these cancer subtypes in other studies.

One new study of U.S. Vietnam veterans who attended a Veterans Affair (VA) Medical Center examined hepatocellular carcinoma comorbid with dual diagnoses of cirrhosis and hepatitis C virus and exposure to Agent Orange (determined by self-report) (Krishnamurthy et al., 2016) [28]. Although the estimated risk was elevated, no statistically significant association was found between exposure to Agent Orange and hepatocellular carcinoma. Such null findings are consistent with studies of other cohorts of U.S. Vietnam veterans as well as those from Australia and Korea. Although the study of Krishnamurthy et al. (2016) reported modest evidence of excess liver cancer among Vietnam veterans using VA Medical Center services, the weak design, nonspecific exposure, and confounding factors remain a concern for interpreting its results. The lack of evidence of an association between exposure and hepatobiliary cancers in the well-designed and exposure-characterized occupational studies does not support an association. Despite the evidence of TCDD activity as a hepatocarcinogen in animals, the evidence from epidemiologic studies remains inadequate to link dioxins with hepatobiliary cancers, which have a relatively low incidence in Western populations. Overall, the available evidence does not support an association between dioxins and hepatobiliary cancers.

Ovadia et al. (2015) [29] conducted an analysis of the relationship between self-reported Agent Orange exposure and long-term outcomes among prostate cancer patients. Data for this analysis were drawn from the Shared Equal Access Regional Cancer Hospital database of 1882 men undergoing radical prostatectomy for prostate cancer between 1988 and 2011 at six VA health care facilities; 333 men (17.7%) were considered to be Agent Orange-exposed. The clinical outcomes reported included the pathologic Gleason Score, pathologic stage, and postoperative pathologic and treatment characteristics as well as prostate cancer-specific death. Cox proportional hazards regression modeling was used to assess the relationship between exposure to Agent Orange and biochemical recurrence, secondary treatment, metastases, and prostate cancer-specific mortality. Models were adjusted for age, race, clinical stage, PSA level, BMI, center, and biopsy Gleason sum. Agent Orange exposure was not found to be associated with biochemical recurrence (HR = 1.21, 95% CI 0.99–1.49), secondary treatment (HR = 1.21, 95% CI 0.97–1.5), metastases (HR = 0.93,

95% CI 0.30–2.66), or prostate cancer-specific mortality (HR = 0.89, 95% CI 0.46–1.85). The study was generally well-conducted, with excellent clinical data and follow-up available within the VA system. Although Agent Orange exposure included an additional level of service location verification to self-report, this measure is still only a proxy for actual initial and subsequent exposure levels. However, a previous study of 93 men by Li et al. (2013) [30] showed good correlation between self-reported Agent Orange exposure and dioxin TEQ levels in adipose tissue. The study's negative results may be applicable to the relationship between Agent Orange and prostate cancer progression but do not directly address initiation and incidence.

In contrast to the results of liver and prostate cancer epidemiological analysis, studies of Vietnam veterans are largely suggestive of modest associations between dioxins exposure and lung cancer incidence and mortality. In studies of U.S. veterans, a significantly increased risk of lung-cancer was found in Army Chemical Corps (ACC) veterans who used herbicides in Vietnam (Cypel and Kang, 2010) [31]), and an increased risk of lung cancer was associated with increased serum TCDD concentrations in Ranch Hand veterans (Pavuk et al., 2005) [8]. The Australian cohort studies of Vietnam veterans (ADVA, 2005) [32], which presumably cover a large portion of exposed soldiers, showed a higher than expected incidence of and mortality from lung cancer. The main limitations of the Australian and American ACC studies are that there was no assessment of exposure and that some potential confounding variables, notably smoking, could not be accounted for. Additionally, the Korean Veterans Health Study (Yi et al., 2014) [33] found modestly elevated, but not statistically significant, relative risks of both lung cancer incidence and mortality compared with the general population and high- versus low-exposure groups. The results were not adjusted for smoking, but earlier self-reported information from a large portion of the cohort indicated that smoking behavior did not appear to be related to the extent of a veteran's exposure to herbicides.

The limited evidence of an association between dioxins and cancer is most likely due to a lower amount of dioxins in these studied populations. The amount of dioxins in some hotspots of Vietnam is large enough to increase their concentration in the blood of exposed people, leading to markedly increased risk ratios. This finding corresponds with the conclusion from the study conducted by Xu et al. [14], which showed that the morbidity and mortality of some types of cancers were related to an increased dioxin concentration in the environment and in the patient's blood. Therefore, it is important to analyze dioxin levels in the contaminated environment and in biological organisms when studying cancer incidence in a population exposed to dioxin.

We planned to conduct a meta-analysis from epidemiological dioxin studies in Vietnamese people but could not complete one because of the limited and heterogeneous data. There was no information about gender or other cofactors in the study by Le et al. [25], which may influence the reliability of the study's results. Therefore, the study cannot be used for further meta-analysis. Nguyen et al. [27] conducted an epidemiological study with a small number of participants (n = 1675), which may have made their results less representative. The study by Do et al. [26] was based on questionnaire data and that lacks information about objective assessment (blood and tissue levels of dioxin). These limitations may reduce the reliability of the obtained results, and further studies are needed to resolve these problems.

5. Conclusions

Based on the literature review of the relationship between dioxins and cancer incidence, the following conclusions can be proposed:

1. The results of epidemiological studies confirm the association between exposure to dioxins and the development of cancer subtypes (soft tissue sarcoma; Hodgkin's and non-Hodgkin's lymphoma; and lung, prostate, and liver cancer) in some contaminated areas of Vietnam.

2. The incidence of these cancer subtypes in Vietnam veterans affected by Agent Orange is much higher than the results of studies published in other parts of the world because of the high levels of dioxins in South Vietnam, which is where Agent Orange was sprayed during the war.
3. In the future, epidemiological studies on the relationship between dioxins and cancers in Vietnam should be conducted using improved research methods to correct the methodological problems of previous Vietnamese studies. In addition, the reliability of the research results needs to be achieved by assessing TCDD exposure level with a large number of study participants.

Funding: This research received no external funding.

Institutional Review Board Statement: The study was conducted according to the guidelines of the Declaration of Helsinki and was approved by the Institutional Review Board of the Vietnam Association for Victims of Agent Orange/dioxin—VAVA (published in Proceeding of International Conference Protocol on 19 December 2021).

Informed Consent Statement: Written informed consent was obtained from the people related to the paper.

Data Availability Statement: Data supporting reported results can be found in Proceeding of International Conference Protocol on 19 December 2021, published by Vietnam Association for Victims of Agent Orange/dioxin—VAVA.

Acknowledgments: I would like to thank Xu, Le, Do, Nguyen and all of the authors of the published papers involved in my study.

Conflicts of Interest: The author declares no conflict of interest.

References

1. IARC Working Group on the Evaluation of Carcinogenic Risks to Humans. Polychlorinated Dibenzo-Para-Dioxins and Polychlorinated Dibenzofurans. In *IARC monographs On the Evaluation of Carcinogenic risks to Humans*; International Agency for Research on Cancer: Lyon, France, 1997; Volume 69, pp. 389–433.
2. IARC Working Group on the Evaluation of Carcinogenic Risks to Humans. Chemical agents and related occupations. In *IARC Monographs on the Evaluation of the Carcinogenic Risk of Chemicals to Human*; International Agency for Research on Cancer: Lyon, France, 2012; Volume 100F, pp. 339–378.
3. Pesatori, A.C.; Consonni, D.; Rubagotti, M.; Grillo, P.; Bertazzi, P.A. Cancer incidence in the population exposed to dioxin after the Seveso accident: Twenty years of follow-up. *Environ. Health A Glob. Access Sci. Source* **2009**, *8*, 39. [CrossRef] [PubMed]
4. Pirkle, J.L.; Wolfe, W.H.; Patterson, D.G.; Needham, L.L.; Michalek, J.E.; Miner, J.C.; Peterson, M.R.; Phillips, D.L. Estimates of the half-life of 2,3,7,8-tetrachlorodibenzo-p-dioxin in Vietnam Veterans of Operation Ranch Hand. *J. Toxicol. Environ. Health* **1989**, *27*, 165–171. [CrossRef] [PubMed]
5. Steenland, K.; Piacitelli, L.; Fingerhut, M.; Chang, L.I. Cancer, heart disease, and diabetes in workers exposed to 2,3,7,8-tetrachlorodibenzo-p-dioxin. *J. Natl. Cancer Inst.* **1999**, *91*, 779–786. [CrossRef] [PubMed]
6. Bodner, K.M.; Collins, J.J.; Bloemen, L.; Carson, M. Cancer risk for chemical workers exposed to 2,3,7,8-tetrachlorodibenzop-dioxin. *Occup. Environ. Med.* **2003**, *60*, 672–675. [CrossRef]
7. Michalek, J.E.; Wolfe, W.H.; Miner, J.C. Health status of Air Force veterans occupationally exposed to herbicides in Vietnam. II. Mortality. *JAMA* **1990**, *264*, 1832–1836. [CrossRef]
8. Pavuk, M.; Michalek, J.; Schecter, A.; Ketchum, N.S.; Akhtar, F.Z.; Fox, K.A. Did TCDD exposure or service in Southeast Asia increase the risk of cancer in air force Vietnam veterans who did not spray agent orange? *J. Occup. Environ. Med.* **2005**, *47*, 335–342. [CrossRef]
9. Collins, J.J.; Bodner, K.; Aylward, L.L.; Wilken, M.; Swaen, G.; Budinsky, R.; Rowlands, C.; Bodnar, C.M. Mortality rates among workers exposed to dioxins in the manufacture of pentachlorophenol. *J. Occup. Environ. Med.* **2009**, *51*, 1212–1219. [CrossRef]
10. Manuwald, U.; Velasco Garrido, M.; Berger, J.; Manz, A.; Baur, X. Mortality study of chemical workers exposed to dioxins: Followup 23 years after chemical plant closure. *J. Occup. Environ. Med.* **2012**, *69*, 636–642. [CrossRef]
11. Ott, M.G.; Zober, A. Cause specifc mortality and cancer incidence among employees exposed to 2,3,7,8-TCDD after a 1953 reactor accident. *J. Occup. Environ. Med.* **1996**, *53*, 606–612. [CrossRef]
12. Warner, M.; Mocarelli, P.; Samuels, S.; Needham, L.; Brambilla, P.; Eskenazi, B. Dioxin exposure and cancer risk in the Seveso Women's Health Study. *J. Environ. Health Perspect.* **2011**, *119*, 1700–1705. [CrossRef]

13. Kogevinas, M.; Becher, H.; Benn, T.; Bertazzi, P.A.; Boffetta, P.; Bueno-de-Mesquita, H.B.; Coggon, D.; Colin, D.; Flesch-Janys, D.; Fingerhut, M.; et al. Cancer mortality in workers exposed to phenoxy herbicides, chlorophenols, and dioxins. An expanded and updated international cohort study. *Am. J. Epidemiol.* **1997**, *145*, 1061–1075. [CrossRef] [PubMed]
14. Xu, J.; Ye, Y.; Huang, F.; Chen, H.; Wu, H.; Huang, J.; Hu, J.; Xia, D.; Wu, Y. Association between dioxin and cancer incidence and mortality: A meta-analysis. *Sci. Rep.* **2016**, *6*, 38012. [CrossRef] [PubMed]
15. Kogevinas, M.; Saracci, R.; Winkelmann, R.; Johnson, E.S.; Bertazzi, P.A.; de Mesquita, B.H.B.; Kauppinen, T.; Littorin, M.; Lynge, E.; Neuberger, M.; et al. Cancer incidence and mortality in women occupationally exposed to chlorophenoxy herbicides, chlorophenols and dioxins. *Cancer Causes Control CCC* **1993**, *4*, 547–553. [CrossRef]
16. Read, D.; Wright, C.; Weinstein, P.; Borman, B. Cancer incidence and mortality in a New Zealand community potentially exposed to 2,3,7,8-tetrachlorodibenzo-p-dioxin from 2,4,5-trichlorophenoxyacetic acid manufacture. *Aust. N. Z. J. Public Health* **2007**, *31*, 13–18. [PubMed]
17. National Academies of Sciences, Engineering, and Medicine. *Veterans and Agent Orange: Update 11*; The National Academies Press: Washington, DC, USA, 2018.
18. The 2020 Issue of the Annual Agent Orange Newsletter (Published Online by US.VA). Available online: https://www.publichealth.va.gov/docs/agentorange/reviews/ao-newsletter-2020.pdf (accessed on 30 May 2022).
19. Kahn, P.C.; Gochfeld, M.; Nygren, M.; Hansson, M.; Rappe, C.; Velez, H.; Ghent-Guenther, T.; Wilson, W.P. Dioxins and dibenzofurans in blood and adipose tissue of Agent Orange-exposed Vietnam veterans and matched controls. *JAMA* **1988**, *259*, 1661–1667. [CrossRef]
20. Schecter, A.; Ryan, J.; Constable, J.D.; Baughman, R.; Bangert, J.; Fürst, P.; Richard, K.W.; Oates, P. Partitioning of 2,3,7,8-chlorinated dibenzo-p-dioxins and dibenzofurans between adipose tissue and plasma lipid of 20 Massachusetts Vietnam veterans. *Chemosphere* **1990**, *20*, 951–958. [CrossRef]
21. Grassman, J.; Cybulska, L. Using a One Health Approach to Understand the Extent of Tcdd Contamination in Vietnam. In Proceedings of the International Conference Diseases Associated with Agent Dioxin/Dioxin Exposure in Vietnam—Current Situation and Solution to Improve Effectiveness of Prevention and Treatment, Hanoi, Vietnam, 19 December 2021; pp. 265–272.
22. Office of the national steering committee 33 ministry of natural resources and environment of Vietnam and Hatfield Consultants. *Environmental and Human Health Assessment of Dioxin Contamination at Bien Hoa Airbase, Viet Nam*; Final Report; Hatfield Consultants: North Vancouver, BC, Canada, 2011.
23. Lorber, M.; Patterson, D.; Huwe, J.; Kahn, H. Evaluation of background exposures of Americans to dioxin-like compounds in the 1990s and the 2000s. *Chemosphere* **2009**, *77*, 640–651. [CrossRef]
24. Goldman, L.R.; Harnly, M.; Flattery, J.; Patterson, D.G., Jr.; Needham, L.L. Serum polychlorinated dibenzo-p-dioxins and polychlorinated dibenzofurans among people eating contaminated home-produced eggs and beef. *Environ Health Perspect.* **2000**, *108*, 13–19. [CrossRef]
25. Le, B.Q.; Doan, H.H. Research on diseases related to toxic chemicals/dioxins in Vietnam veterans. *Mil. Med.* **2007**, *32*, 69–73.
26. Do, Q.T. *Agent Orange and the Prevalence of Cancer among the Vietnamese Population 30 Years after the End of the Vietnam War*; The World Bank Research Working Paper; World Bank: Washington, DC, USA, 2009.
27. Nguyen, T.N.P.; Pham, D.T. Basic Investigating the Health Status of Victims from Agent Dioxin in Some Provinces at South of Vietnam. In Proceedings of the International Conference Diseases Associated with Agent Dioxin/Dioxin Exposure in Vietnam—Current Situation and Solution to Improve Effectiveness of Prevention and Treatment, Hanoi, Vietnam, 19 December 2021; pp. 28–45.
28. Krishnamurthy, P.; Hazratjee, N.; Opris, D.; Agrawal, S.; Markert, R. Is exposure to Agent Orange a risk factor for hepatocellular cancer? A single-center retrospective study in the U.S. veteran population. *J. Gastrointest. Oncol.* **2016**, *7*, 426–432. [CrossRef]
29. Ovadia, A.E.; Terris, M.K.; Aronson, W.J.; Kane, C.J.; Amling, C.L.; Cooperberg, M.R.; Freedland, S.J.; Abern, M.R. Agent Orange and long-term outcomes after radical prostatectomy. *Urol. Oncol.* **2015**, *33*, 329. [CrossRef] [PubMed]
30. Li, Q.; Lan, L.; Klaassen, Z.; Shah, S.R.; Moses, K.A.; Terris, M.K. High level of dioxin-TEQ in tissue is associated with Agent Orange exposure but not with biochemical recurrence after radical prostatectomy. *Prostate Cancer Prostatic Dis.* **2013**, *16*, 376–381. [CrossRef] [PubMed]
31. Cypel, Y.; Kang, H. Mortality patterns of Army Chemical Corps veterans who were occupationally exposed to herbicides in Vietnam. *Ann. Epidemiol.* **2010**, *20*, 339–346. [CrossRef] [PubMed]
32. Wilson, E.; Horsley, K. *Australian Vietnam Veterans Mortality Study 2005*; Department of Veterans' Affairs: Canberra, Australia, 2005.
33. Yi, S.W.; Ryu, S.Y. Agent Orange exposure and risk of death in Korean Vietnam veterans: Korean Veterans Health Study. *Int. J. Epidemiol.* **2014**, *43*, 1825–1834. [CrossRef]

MDPI
St. Alban-Anlage 66
4052 Basel
Switzerland
Tel. +41 61 683 77 34
Fax +41 61 302 89 18
www.mdpi.com

Toxics Editorial Office
E-mail: toxics@mdpi.com
www.mdpi.com/journal/toxics